Thin Film Solar Cells: Principles and Applications

Thin Film Solar Cells: Principles and Applications

First published June 2016
Written by Jin Hyeok Kim (Ph.d), Chonnam National University
Publisher: Byung-Moon Jee
ISBN 978-89-6849-324-9 (93560)
Price: 13,000 Won

Published by Chonnam National University Press,
77 Yongbong-ro, Bukgu, Gwangju,
Republic of Korea, 61186
Tel: 82-62-530-0571/3, Fax: 82-62-530-0579

Printed and bound in the Republic of Korea

CIP No : CIP2016016851

Thin Film Solar Cells: Principles and Applications

Jin Hyeok Kim

R. B. V. Chalapathy, Archana S. Kamble,

Sachin A. Pawar, Mahesh P. Suryawanshi,

Abhishek C. Lokhande

Chonnam National University, Korea

Chonnam National University PRESS

Preface

Today world population is 7 billion on the planet earth. In 2,100, the expected population is estimated to increase up to 10 billion. To meet the basic necessities for growing population enormous pressure has been induced on the available energy sources. So far, the energy is met by burning fossil fuels such as wood, coal, oil and natural gas. By consuming these energy sources there are adverse effects on the environment. Pollution and global warming due to these energy sources has already got a hold of the earth, which directly and indirectly affects the human life. Hence, renewable energy sources such as wind energy, biomass energy, geothermal, photovoltaics were developed as energy alternatives for non-renewable energy sources.

Amongst these, photovoltaics which rely on solar energy is the best substitute due to extensive, inexhaustible, clean and pollution-free. The lack of wide spread use of photovoltaics is due to expensive compared to conventional energy sources. The continuous development of technologies for thin-film solar cells has reduced the price of solar cells in recent years. Thin-film solar cells were considered as low-cost compared to crystalline and wafer-based photovoltaics. Hence, further development of materials with less toxic, most abundant elements and large area production are necessary to mitigate the materials limitation due to wide use of thin film solar cells.

Keeping in mind that photovotaics is the preferred choice of energy for the coming next 50-100 years, the photovoltaic science and technology is continuously evolving. In order to keep the development, it is necessary to train the next generation students in the field of photovoltaics. The book is written with a focus on the undergraduate and master's students to introduce the concepts of thin-film solar cells. The book has been organized into 7 chapters.

In chapter 1, the fundamental concepts of different types of energy sources were introduced. In chapter 2, the basic concepts of solar energy and solar insolation and air mass were introduced. Keeping the view that materials used in the photovoltaics, the properties of basic semiconductors were described in the third chapter. The fourth chapter explains the p-n junction and solar cells. In 5th chapter, the fabrication of CIGS, CdTe and CZTSSe and CTS thin-film solar cells were explained. In 6th chapter, various characterization techniques used for solar cells analysis are given. Finally, chapter 7 describes the various strategies for further improvement in the power conversion efficiency of thin-film solar cell. This book meets one semester course on thin film photovoltaics for undergraduate students.

I am thankful to my students Dr. R. B. V. Chalapathy, Dr. Archana Kamble, Dr. Sachin Pawar, Dr. Mahesh Suryawanshi and Mr. Abhishek Lokhande for helping whole heartedly to complete this book. I am thankful to my master's and graduate students, post docs and visiting scientists in Photonic and Electronic Thin Film Laboratory and Optoelectronics convergence research center in the Department of Materials Science and Engineering, Chonnam National University who worked on various thin film solar cells and raised many topics related to solar cells, processing and research problems. I am grateful to the Human Resources Development of the Korea Institute of Energy Technology Evaluation and Planning (KETEP) Grant funded by the Korea government Ministry of Trade, Industry and Energy (GET-FUTURE). My thanks are also due to the members of Chonnam National University press for their help in bringing this manuscript in a book form. Finally, I am grateful to my wife and children for allowing me spare time to write this book and their constant encouragement without their support this book is not in this form.

This book is dedicated to all the persons who work and sacrifice their time and energy for developing the green technologies for the well-being of humanity.

Kim, Jin Hyeok

Gwangju, South Korea, May 2016

Contents

Symbols and Acronyms

A	area
C	capacitance
c	is velocity of light
D_n	diffusion coefficient of electrons
D_p	diffusion coefficient of holes
e	electronic charge
E_A	energy of the acceptor
Ea	activation energy
E_c	energy of conduction band energy
E_D	energy of the donor
E_F	Fermi energy
E_{Fi}	intrinsic Fermi energy
E_g	energy band gap
E_i	energy of intrinsic level
E_t	energy of traps
E_v	energy of valence band energy
$F(E)$	Fermi energy
FF	fill factor
G_n	generation rate of electrons
G_p	generation rate of holes
h	is Planck's constant

I	excitation intensity
I_L	light generated current
I_{mp}	current maximum
I_o	dark current density
IQE	internal quantum efficiency
I_{sc}	short circuit current
j	current density
J_n	current density due to electrons
J_p	current density due to holes
J_{sc}	short circuit current density
k_b	Boltzman constant
L_n	diffusion length of electrons
L_p	diffusion length of holes
m_e	effective mass of electron
m_h	effective mass of hole
n	electrons
N_a	acceptor density
N_A	Avogadro's number
N_c	effective conduction band density of states
N_d	donor density
n_i	intrinsic carrier density
N_{st}	density of surface trap states
N_t	density of trap states
N_v	effective valence band density of states
P	excitation power
p	holes
P_{in}	incident power

q	electronic charge
QE	quantum efficiency
R_n	recombination rate of electrons
R_p	recombination rate of holes
R_s	series resistance
R_s	sheet resistance
R_{sh}	shunt resitance
T	Temperature
t	thickness
U_n	net recombination of electrons for unit volume
U_p	net recombination of holes for unit volume
V_d	drift velocity
V_{mp}	voltage maximum
V_{oc}	open circuit voltage
V_T	thermal voltage
v_{th}	thermal velocity
α	absorption coefficient
ε	permittivity
ε_o	vacuum permittivity
η	conversion efficiency
λ	wavelength of light
μ	mobility
μ_n, μ_p	mobility of electron; hole mobility
ρ	resistivity
σ	conductivity
τ_n, τ_p	life time of electron; hole life time

Acronyms

AM	air mass
AZO	aluminum doped zinc oxide
CdS	Cadmium sulfide
CdTe	cadmium telluride
CIGS	copper indium gallium selenide
CSP	(Concentrated Solar Power)
CST	(Concentrated Solar Thermal)
CTS	copper tin sulfide
CZTS	Copper zinc tin sulfide
CZTSe	copper zinc tin selenide
CZTSSe	copper zinc tin sulfo-selenide
DLCP	drive level capacitance profile
DLTS	deep level transient spectroscopy
EL	electroluminescence
EQE	external quantum efficiency
FTO	fluorine doped tin oxide
i-ZnO	intrinsic zinc oxide
EIA	(Energy Information Agency)
Mo	molybdenum
PCE	power conversion efficiency
PL	photoluminescence
SRH	Shockely-Read-Hall recombination
TCO	transparent conducting oxide

Chapter 1
Energy demand and its resources

1.1. Introduction

A large number of energy sources are available to human being on this earth to provide their day-by-day energy needs. However, the use of particular energy source should be made in each case on the basis of economic, environmental and safety considerations. In this chapter, we will consider briefly the world energy consumption that depends on these number of available energy resources, the use of these energy sources to provide our daily needs, the problems associated with these energy sources and some several scenarios for future.

1.2. Energy resources

There are two types of energy sources available to mankind, which can be categories as: (1) non-renewable energy sources, i.e. those sources of energy which cannot be replaced or produced on any time scale less than million years once used; and (2) renewable energy sources, i.e. those sources of energy which are available naturally for millions of years. Table 1.1 shows a listing of these energy sources according to their category.

In Table 1.1, the energy sources are listed as per their use by mankind to provide its daily uses. The non-renewable energy sources, which are the prime sources of energy for mankind, will eventually become exhausted, while the renewable energy sources will not. Therefore, we should consider the global energy consumption in order to know how long will be the non-renewable energy sources last.

Table 1.1. Different types of available energy sources to mankind

Non-renewable energy sources	Renewable energy sources
Fossils fuels (coal, oil, natural gas)	Biological sources (wood, plants)
Geothermal	Hydropower (dams, tides)
Nuclear fission	Wind energy
Nuclear fusion	Solar energy

1.3. Energy economy

1.3.1. Global energy consumption

Global energy consumption has about doubled in the last three decades of the past century. The most recent estimate of the world energy consumption was 5.67×10^{20} joules, or 157,481 TWh, in 2013. According to the Energy Information Administration (EIA), the total world energy consumption was 143,851 TWh in 2008, 133,602 TWh in 2005, 117,687 TWh in 2000, and 102,569 TWh in 1990. In 2013, about 81.4% of the primary energy consumption was from fossil fuels (oil 31.1%, coal 28.9%, natural gas 21.4%), 4.8% from nuclear fuels, 10.2% from biofuels and waste and the remaining 3.6% from renewable energy sources (out of which 2.4% from hydro and 1.2% from 'other' (solar, wind, geothermal, heat, etc.)). Oil, coal, and natural gas were the most popular energy fuels [1]. The world demand for energy is expected to increase steadily until 2030 according to many scenarios. Global primary energy demand is projected to increase by 1.7% per year from 2000 to 2030, reaching an annual level of 15.3×10^9 tons of oil equivalent (toe). The projected growth is, nevertheless, slower than the growth over the past 30 years, which ran at 2.1% per year. The global oil demand is expected to increase by about 1.6% per year from 75×10^6 barrels per day to 120×10^6 barrels per day [2]. The steady increase in world's population and to strive for better development and comfort are the two main reasons for the increase in the energy consumption at all times. The world population is expected to almost double in the next 50 years. Thus, energy demand could double or triple as population increases and developing countries expand their economies and overcome poverty by 2050. With this increase in the population, the per capita consumption of energy is also increasing. This rise in population as well as in the economy are the two main parameters that will cause to increase the energy demand in the coming decades. With such increasing energy demand, it is accepted that there will be energy crises in the near future due to limited amount of readily available fossils fuels.

1.4. Fossils fuels and its impact on environment

The fossil fuels such as coal, oil, and natural gas (also known as non-renewable energy source), are limited in reserves and it is predicted that the demand will soon exceed the supply based on the continues increasing energy consumption. Therefore, it is necessary to reduce their exploitation rate by partial replacements, especially through the use of renewable energy sources. In addition, the increased use of these non-renewable energy sources resulted into several negative environmental impacts [3]. Burning of fossils fuels has caused global-scale increase in CO_2 in the atmosphere, which may lead to global climatic change in the near future. The use of fossils fuels in automobiles industries produces exhaustive gases that increase the air pollution as well as the surface ozone concentration, which is very dangerous to human health and environment. Therefore, it is today's need to opt the available renewable energy sources to get rid of or/ reduce these unwanted and damaging effects to environment as well human health.

1.5. Renewable energy sources

1.5.1. Wind energy

Wind is a form of solar energy. The uneven heating of the atmosphere by the sun, the irregularities of the earth's surface, and rotation of the earth causes the wind energy [4]. The flow patterns of wind are modified by the earth's terrain, bodies of water, and vegetative cover. The modern wind turbines, which are used to generate electricity by harvesting this wind flow, or motion energy [4].

1.5.2. How wind power is generated

The kinetic energy of wind can be changed into other forms of energy, either mechanical or electrical energy and this process can be termed as "wind energy" or "wind power". Wind turbine is a device that converts the kinetic energy from the wind into mechanical or electrical power. The mechanical power can be used for specific tasks (such as grinding grain or pumping water).

1.5.3. Wind turbine types

Horizontal axis design Vertical axis design

Fig. 1.1. Photo images of horizontal axis and vertical axis wind turbine designs.

(Sources: https://en.wikipedia.org/wiki/Wind_power and https://en.wikipedia.org/wiki/Troposkein)

Wind turbines power an electric generator by turning in the moving air and deliver an electric current. In simple words, a wind turbine is the opposite of a fan. Wind turbines use wind to make electricity, whereas fan uses electricity to make wind.

Modern wind turbines are categorized into two basic groups; the horizontal-axis design, like the traditional farm wind mills, and the vertical-axis design, like the eggbeater-style Darrieus model, named after its French inventor. Most large modern wind turbines are horizontal-axis turbines.

1.5.4. Turbine components

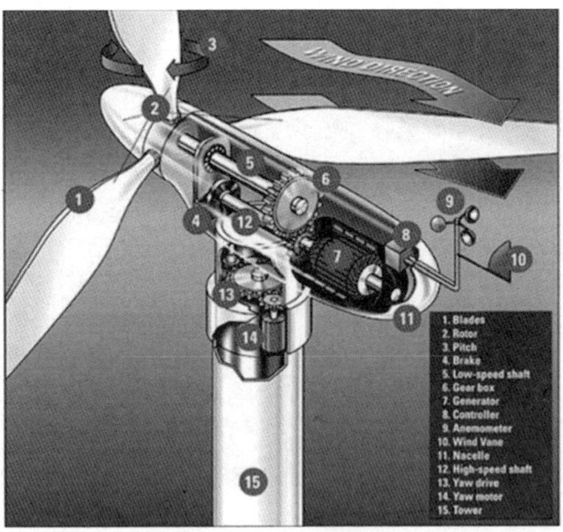

Fig. 1.2. Schematic representation of wind turbine with its each component.

Fig. 1.2 shows the schematic representation of horizontal axis wind turbine with its different components. blade or rotor, which converts the energy in the wind to rotational shaft energy; adrive train, usually including a gearbox and a generator; a tower that supports the rotor and drive train; and other equipment, including controls, electrical cables, ground support equipment, and interconnection equipment.

Wind turbines are often grouped together into a single wind power plant, also known as a wind farm, and generate bulk electrical power. Electricity from these turbines is fed into a utility grid and distributed to customers, just as with conventional power plants.

1.5.5. Wind turbine size and power ratings

Wind turbines are available in a variety of sizes, and therefore power ratings. The largest machine has blades that span more than the length of a football field, stands 20 building stories high, and produces enough electricity to power 1,400 homes. A small home-sized wind machine has rotors between 8 and 25 feet in diameter and stands upwards of 30 feet and can supply the power needs of an all-electric home or small business. Utility-scale turbines range in size from 50 to 750 kilowatts. Single small turbines, below 50 kilowatts, are used for homes, telecommunications dishes, or water pumping.

1.5.6. Solar thermal technologies

Solar thermal energy (STE) is a form of energy and a technology for harnessing solar energy to generate thermal energy or electrical energy for use in industry, and in the residential and commercial sectors. The first installation of solar thermal energy equipment occurred in the Sahara approximately in 1910 by Frank Shuman when a steam engine was run on steam produced by sunlight. Because liquid fuel engines were developed and found more convenient, the Sahara project was abandoned, only to be revisited several decades later [1].

Solar thermal collectors are classified by the United States Energy Information Administration as low-, medium-, or high-temperature collectors. Low-temperature collectors are flat plates generally used to heat swimming pools. Medium-temperature collectors are also usually flat plates but are used for heating water or air for residential and commercial use. High-temperature collectors concentrate sunlight using mirrors or lenses and are generally used for fulfilling heat requirements up to 300 deg ℃ / 20 bar pressure in industries, and for electric power production. Two categories include Concentrated Solar Thermal (CST) for fulfilling heat requirements in industries, and Concentrated Solar Power (CSP) when the heat collected is used for power generation. CST and CSP are not replaceable in terms of application. The largest facilities are located in the American Mojave Desert of California and Nevada. These plants employ a variety of different technologies. The largest examples include, Ivanpah Solar Power Facility (377 MW), Solar Energy Generating Systems installation (354 MW), and Crescent Dunes (110 MW). Spain is the other major developer of solar thermal power plant. The largest examples include, Solnova Solar Power Station (150 MW), the Andasol solar power station (150 MW), and Extresol Solar Power Station (100 MW).

1.5.7. Biomass energy technologies

There are many types of biomass — organic matter such as plants, residue from agriculture and forestry, and the organic component of municipal and industrial wastes — that can now be used to produce fuels, chemicals, and power. Wood has been used to provide heat for thousands of years. This flexibility has resulted in increased use of biomass technologies. According to the Energy Information Administration, 53% of all renewable energy consumed in the United States was biomass-based in 2007.

Fig. 1.3. A cogeneration plant in Metz, France. The station uses waste wood biomass as an energy source, and provides electricity and heat for 30,000 dwellings. (Source: Wikipedia)

Biomass technologies break down organic matter to release stored energy from the sun. The process used depends on the type of biomass and its intended end-use.

1.6. Global warming and climate change

Global warming and climate change are terms for the observed century-scale rise in the average temperature of the Earth's climate changeand its related effects. Global warming and climate change refer to an increase in average global temperatures. Activities including natural events and human activities are responsible to be contributing to an increase in average global temperatures. This is caused primarily by increases in greenhouse gases such as Carbon Dioxide (CO_2). Six main greenhouse gases are carbon dioxide (CO_2), methane (CH_4) and nitrous oxide (N_2O), plus three fluorinated industrial gases: hydrofluorocarbons (HFCs), perfluorocarbons (PFCs) and sulfur hexafluoride (SF_6). Water vapor is also considered a greenhouse gas [1].

Energy from the sun drives the earth's weather and climate, and heats the earth's surface. In turn, the earth radiates energy back into space. Some atmospheric gases (water vapor, carbon dioxide, and other gases) trap some of the outgoing energy, retaining heat somewhat like the glass panels of a greenhouse. These gases are therefore known as greenhouse gases. The greenhouse effect leads to rise in temperature on earth as certain gases in the atmosphere trap energy.

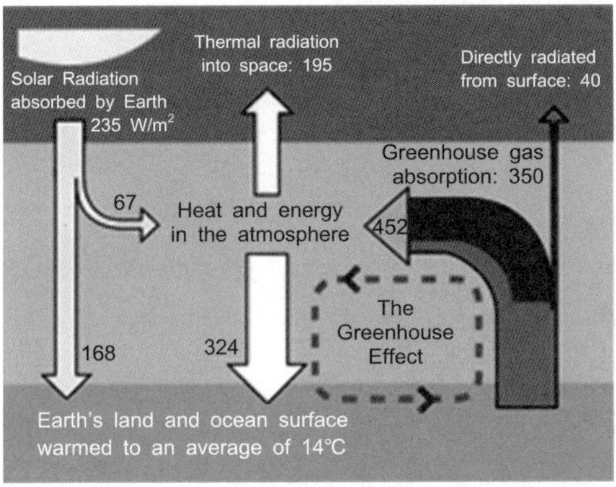

Fig. 1.4. The schematic representation of the greenhouse effect [5].

This increase in global temperature has serious impact on the nature and life on the planet. The rapid melting of ice, increase in the sea water level, hurricanes, precipitation, water shortage and disturbance in ecosystem are the impacts of global warming. Other likely changes include more frequent extreme weather events including heat waves, droughts, heavy rainfall with floods and heavy snowfall, ocean acidification; and species extinctions due to shifting temperature regimes. Fig. 1.5 represents the effects of global warming and climate change. The Intergovernmental Panel on Climate Change (IPCC) reported in 2014 that scientists were more than 95% certain that global warming is mostly being caused by increasing concentrations of greenhouse gases (GHG) and other human (anthropogenic) activities [6]. Hence, it is a necessity to act and find suitable solution for the problem of global warming. Energy sources such as fossil fuels, coal and petroleum products are widely used. However, these energy sources are limited and

Fig. 1.5. The effects of global warming and climate change (source: Wikipedia).

also their consumption cause pollution and global warming effects. Therefore alternatives for these energy sources should be considered.

There is no single solution for global warming. Many solutions must be reasonably adopted for solving this issue. The first and the foremost approach is to use high efficient sources which can reduce cost, pollution and hence money. Another approach is the utilization of vehicle transport systems which have low CO_2 emissions. Renewable energy sources such as solar, wind, geothermal and bioenergy are available around the world, which can be used widely to meet the vast majority of our energy needs. Renewable technologies can be employed quickly, are increasingly cost-effective and pollution free. Another approach by which global warming can be fought by planting trees which can reduce emissions. Ensuring sustainable development, developing low and zero carbon technologies can also play a contributory role in reducing global warming. Developed countries must take leadership in providing financial and technical help for fighting global warming issues.

1.7. Solar photovoltaics as solution to energy

Energy crisis have attracted attention of the entire world as the available energy sources are on the route of depletion. Due to the limited availability of these energy sources, their prices are rising very high. Another great concern is the ill effects due to their consumption on the environment. Pollution and global warming due to these energy sources has already disturbed earth's atmosphere which directly and indirectly affects the human life. Hence, substitutes for these energy sources are needed to address the current serious issues. Solar energy is extensive, inexhaustible, clean, free and renewable energy which can be the best substitute. The current research is mostly diverted towards the development of efficient use of solar energy. Many countries have been actively engaged their research, technology and resources for solar initiatives in order to overcome the energy crises problems [7].

Photovoltaic (PV) system is totally based on the use of solar energy. It employs a device (solar cell) which does the conversion of light to electricity. Today, the electricity generated from solar cell is nearly 1% of the total worldwide electricity production. Realizing the rapid development in the PV technology, the Joint Research Centre (Europe) anticipated 20% energy consumption from sunlight and will increase up to 50% in the next 60 years. Therefore, the reliability on the photovoltaic for satisfying the energy needs is increased and hence efforts are invested to produce solar cells with high power conversion efficiencies (PCE). First generation solar cells perform very well at ideal conditions (25℃ and light intensity 1,000 W/M^2). These solar cells absorb light in the optical range of 400 to 1,000 nm. Their optical absorbance is poor in UV and IR region. Hence, in low and diffused light conditions they perform poorly. However, second generation thin film solar cells have good performance in diffused light conditions as they have good absorbance in the UV visible and IR region. Hence, compared to first generation solar cells, second generation solar cells are cheap and have better performance in condition of diffused light. The second generation solar cells based on CdTe and CIGS absorber compound have also showed the highest PCE upto 21.5% and 21.7%, respectively. However, these absorber compounds cannot be used as they employ toxic cadmium and rare earth tellurium and indium compounds despite they have shown promising efficiency. Hence, search for an absorber compound composed of earth abundant and nontoxic elements is considered [8].

References

1. "Key World Energy Statistics 2015", http://www.iea.org/publications/freepublications/., IEA. 2015. pp.8-37.
2. Zaki Sen, Solar energy fundamentals and modeling techniques, Springer-Verlag London Limited, (2008).
3. P. D. Dunn, Renewable energies: sources, conversion and application, Peregrinus, Cambridge, (1986).
4. S. C. Bhatia, Advanced renewable energy systems (Part 1 & 2), Woodhead Publishing India in Energy, (2014).
5. http://www.globalissues.org/article/233/climate-change-and-global-warming-introduction.
6. https://en.wikipedia.org/wiki/global_warming
7. A. C. Lokhande, R. B. V. Chalapathy, Mingrui He, Eunjin Jo, Myengil Gang, S. A. Pawar, C. D. Lokhande, Jin Hyeok Kim, Development of Cu_2SnS_3 (CTS) thin film solar cells by physical techniques: A status review, Solar Energy Materials & Solar Cells, 153, 8410 (2016).
8. S. R. Taylor, S. M. McLennan, The Continental Crust: Its Composition and Evolution, Blackwell Scientific Publication, Oxford, (1985), pp.1-312.

Chapter 2
Solar energy for electricity

2.1. Introduction

As we have seen in previous Chapter 1 that the majority of world's energy demands is supplied from fossil fuels such as coal, oil and natural gas, which are non-renewable energy sources and are going to be depleted one day. In addition, these non-renewable energy sources are facing enormous challenges such as rising prices and increasing the environmental concerns over the sudden climate change due to increased demand of these non-renewable energy sources. Therefore, the efforts are underway from all over the world to increasingly support the development of alternative renewable energy sources. Variety of renewable energy sources such as wind, biomass, geothermal, hydroelectric and solar energy, which are generally available unlimited, have emerged as potential energy alternatives to non-renewable energy sources. Among the different renewable energy sources, solar energy has emerged as one of most rapidly growing renewable energy sources of electricity generation. Therefore, everyone should understand why solar energy is important renewable energy source among the different renewable energy sources.

2.2. Why solar energy?

Solar energy is ultra clean, natural, easily available, free of cost and sustainable source of energy that can be utilized in the solar electricity generation, solar heating and cooling appliances as well as solar lighting appliances. There are several advantages of solar energy for electricity generation. It does not require the use of fossil fuels for electricity generation and there for it is independent of the limited reserves of natural sources. It produces the electricity with no/or limited impact on the environment. It can be used in the remote areas effectively, where it is too expensive to extend the electricity power grid [1]. Many everyday appliances such as calculators and low power consuming devices can be powered by solar energy effectively. However, solar energy has also some disadvantages such as it can only be available in sunny day time. The initial capital investment in purchasing and installing solar collectors, panel and cells is relatively expensive. Since only 40% of the sunlight is absorbed by the solar panels, the remaining 60% of the sunlight is wasted and is not absorbed. Although, new technologies offer 80% of the sunlight to get absorbed, however the cost of these solar panels is very high [2]. Unlike other renewable energy sources which can be operated at night time as well, solar energy is not available at night, so ultimately we have to depend on the local power grid. Apart from that the winter, rainy seasons as well storm and hurricanes reduce the ability to draw the energy from sunlight. Despite of such disadvantages, solar energy being ultra clean, natural, easily available, free of cost and sustainable source of energy has been recognized as a major sources of energy for many years.

2.3. Origin of solar energy

Before being proceeded further, we first must get acquitted with the origin of this solar energy, which is considered to be a major source of renewable energy sources. As an account for world's energy need and demand are incomplete without the discussion of the sun. The sun is origin of solar energy. The sun is a dG2 star, classified as a yellow dwarf of the fifth magnitude. The sun has a mass of approximately 10^{24} tons, a diameter of 1.39 x 10^6 km, and radiates energy at a rate of some 3.8 x 10^{20} megawatts and it is predicted that it will continue for several billion years in the future. It is an internal energy generator and distributor to other planets such as the earth. The core temperature of the sun varies between 8 x 10^6 and 40 x 10^6 K and the density is estimated at about 100 times that of water. The distance between the sun and the earth varies from 92 to 95 million miles. The variation in the distance is because of the orbit of the earth around the sun, which is an ellipse. The inverse square law states that the light energy reaching the earth surface in June (when the earth is at its maximum distance from the sun) is approximately 94% of the light energy reaching the earth in December [3].

The interior of the sun is a nuclear furnace that releases energy due to nuclear fusion reactions at the sun's core in much the same way as man-made thermonuclear explosions. The temperature at the sun's core is high enough to facilitate these nuclear fusion reactions, which are assumed to be the source of the sun's energy. The temperature of sun's core is of the order of the 10^6 degrees, whereas the temperature of the surface layer of the sun, so called as photosphere, is about 6000 K and closely approximates a blackbody. The total power density of the solar radiation at the mean earth-sun distance on a plane perpendicular to the direction of the sun, outside the earth's atmosphere, is referred to as the ***solar constant***. This quantity is difficult of measure from the surface to the earth due to the effect of the atmosphere.

Table 2.1. Chemical composition of the sun

Element	Atomic number	Molecular weight	Abundance (% of Mass)
Hydrogen	1	1.008	71.0
Helium	2	4.003	27.1
Oxygen	8	16.000	0.97
Carbon	6	12.011	0.40
Nitrogen	7	14.007	0.096
Silicon	14	28.086	0.099
Magnesium	12	24.312	0.076
Neon	10	20.183	0.058
Iron	26	55.847	0.14

Langley in 1881 for the first put a method for the effective measurement of solar constant [4-6]. This value varies slightly throughout year due to the variation of the distance between the sun and the earth. This value is maximum of about 1,400 W/m^2 on around January 3, when the sun is closest to the earth and when the sun is farthest away from the earth, i. e. on July 4, it is 1,353 W/m^2 [7]. This power density is available, on the sunlit side of the earth, 24 hours a day, each day of the year, yielding an annual energy flux to the earth of 1,186 kWh/cm^2. The total power that is incident on the earth's surface from the sun every year is 1.73 x 10^{14} kW and this is equivalent to 1.5 x 10^8 kWh annually [8]. If we compare this to naturally available non-renewable energy source (coal), then it is equivalent to 1.9 x 10^{14} coal tons. Table 2.1 shows the chemical composition of the sun. It is seen that the hydrogen and helium are the most abundant elements in the sun. The second most abundant element i.e. helium was discovered from the spectrum of solar radiation, which comes from the Greek word "helios" [9].

Properties of sunlight

Sunlight is a form of "electromagnetic (EM) radiation". The EM spectrum is the range of all types of EM radiations. The radiation is energy that travels and spreads out as it goes. The EM spectral components mostly are in the visible, near infrared, and near ultraviolet. The other types of EM radiation that make up the electromagnetic spectrum are microwaves, X-rays and gamma-rays. Fig. 2.1 shows the EM spectrum.

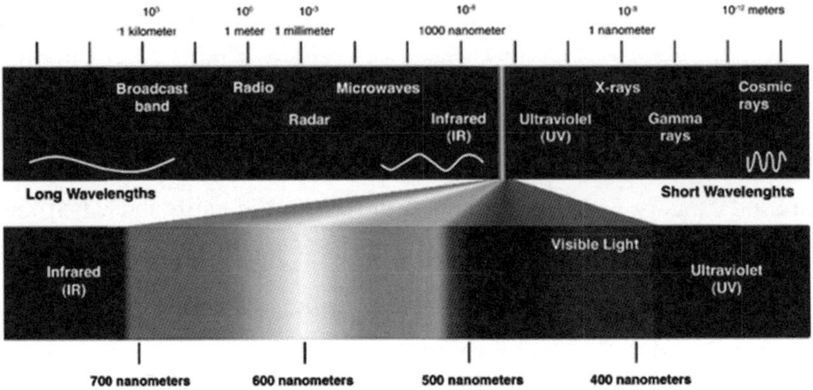

Fig. 2.1. The electromagnetic (EM) spectrum.

(Source: http://www.astronomersgroup.org/EMspectrum.html, Accessed on May 5, 2016 at 11.22 KST)

The visible light was viewed as a small subset of the EM spectrum since the late 1980's. The EM spectrum describes light as continuous number of waves with different wavelengths. The description of light as a wave first gained acceptance in the early 1800's when experiments by Thomas Young, François Arago, and Augustin Jean Fresnel showed interference effects in light beams, indicating that light is made of waves. By the late 1860's light was viewed as part of the EM spectrum. Meanwhile, Plank proposed that the total energy of light is made up of indistinguishable energy elements, or a quanta of energy. Einstein, while examining the photoelectric effect correctly distinguished the values of these quantum energy elements. For their work in this area, Planck and Einstein won the Nobel prize for physics in 1918 and 1921, respectively and based on this work, light may be viewed as consisting of "packets" or particles of energy, called photons.

Nowadays, quantum mechanics explains the wave-particle duality nature of light. However, a complete physical description of the properties of light requires a quantum-mechanical analysis of light, since light is a type of quantum-mechanical particle called a photon. According to the Plank's hypothesis, all electromagnetic radiation is quantized and occurs in finite "bundles" of energy, which is also called as photons. The quantum of energy for a photon is not Planck's constant h itself. However, it is characterized by either a wavelength, denoted by λ or equivalently an energy, denoted by E. There is an inverse relationship between the energy of a photon (E) and the wavelength of the light (λ), which is given by the equation:

$$E = \frac{hc}{\lambda} \tag{2.1}$$

where, h is Planck's constant (h = 6.63 x 10^{-34} joule.s) and c is speed of the light (c = 2.998 x 10^{8} m/s).

2.4. Solar radiation

The power density of solar radiation outside the earth's atmosphere is 1,366 W/m^2, according to several decades of careful measurements [7]. However, the solar radiation is reduced by about 22% on the surface of the earth, when the sun is right to zenith due to scattering and absorption even under a perfectly clear sky. However, the variation in the power density is small and it is considered to be constant for photovoltaic applications. When the solar radiation has reached the earth's surface, the number of additional effects such as atmospheric effects; including absorption and scattering, local variations in the atmosphere; such as water vapor, clouds and dust particles, and the seasons of the year and the time of day plays their part and due to which the solar radiation varies widely while reaching the earth's surface. These all have the general overall effect of reducing the overall power density received at the earth's surface.

Atmospheric attenuation

The solar radiation reaching the earth's surface is reduced due the scattering, reflection and absorption by the atmosphere. A part of the scattered radiations by the molecules or particles in the atmosphere reaches the earth's surface from the entire sky and it is called as "diffuse radiation", whereas the solar radiation that comes directly through the atmosphere is called "direct or beam radiations". The diffused radiation is undirected, and so it seems to just coming from any which way of the region of the sky, whereas the later one is unidirectional and travels all in the same direction. The direct radiation is around 85% of the total insolation receiving the earth's surface and the diffuse radiation is about 15% under the perfectly clear sky and when the sun is very high in the sky. However, the diffuse radiation increases up to 40% when the sky goes lower in the sky, i. e. when the sun is 10° above the horizon [10]. Furthermore, the atmospheric conditions such as clouds and dust particles due to the pollution mentioned above also increases the percentage of diffused radiation and it is almost 100% on an extremely cloudy day. Thus, the degree of attenuation of solar radiation is travelling through earth's atmosphere depends on the length of path and the characteristics of medium in which the radiation is travelling.

Fig. 2.2 shows what happens when the solar radiation passes through the earth's atmosphere. As mentioned above, a part of solar radiation entering the earth's surface gets reflected, scattered or absorbed. Here is the quantitative summary about these phenomena. Approximately, 30% of solar radiation is reflected or scattered back to space.

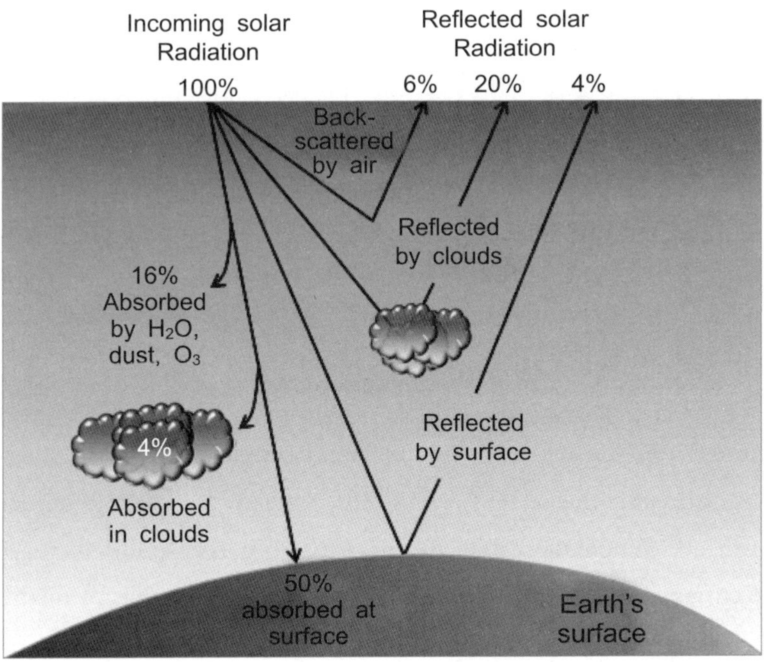

Fig. 2.2. Interaction of solar radiation with atmosphere.

20% of solar radiation reaching the earth's surface is reflected by clouds, whereas 4% is reflected by the surface of Earth and the remaining 6% is scattered by air. Out of remaining 70%, 20% is absorbed by the atmosphere; where the water vapor, dust, and O_3 from the atmosphere absorb 16% and another 4% is absorbed by the clouds. The remaining 50% is absorbed by the solid surface of Earth. Thus, the total radiation energy received by the atmosphere and the solid Earth is about 70% [11].

Air mass

In solar radiation calculations, one standard "air mass" which is defined as the length of path of the atmosphere that has to be passed by the solar radiation, when the sun is at its zenith. It is abbreviated as "AM" and it is ratio between how much atmosphere that the solar radiation has to pass through right that moment and how much atmosphere that the radiation would have pass through if the sun was directly overhead, as shown in Fig. 2.3 [12]. The air mass is defined as:

$$AM = \frac{1}{\cos(\theta)} \tag{2.2}$$

where, θ is the angle from the zenith angle.

The air mass is 1 (AM 1.0), when the sun is directly overhead. When the angle of the sun from the zenith increases, the air mass increases so that at about 48° from the vertical air mass is 1.5 (AM 1.5) and at 60°, it is 2.0 (AM 2.0). It is assumed that the increase in the air mass is assumed to directly proportional to the reduction in the intensity of solar radiation with increasing zenith angle of the sun. Therefore, it varies with the time and day of the year as well as the altitude of the location, making it necessary to choose a specific set of air mass and atmospheric condition for the reference spectra [13].

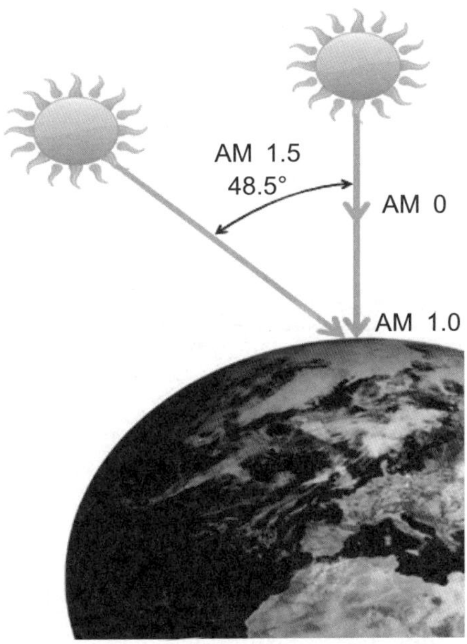

Fig. 2.3. The amount of atmosphere (air mass) through which radiation from the sun must pass to reach the earth's surface depends on the sun's position.

The Air Mass (AM) can be estimated at any location using the following formula:

$$AM = \sqrt{1 + \left(\frac{s}{h}\right)^2} \tag{2.3}$$

where s is the length of the shadow cast by a vertical post of height h, as shown in Fig. 2.4.

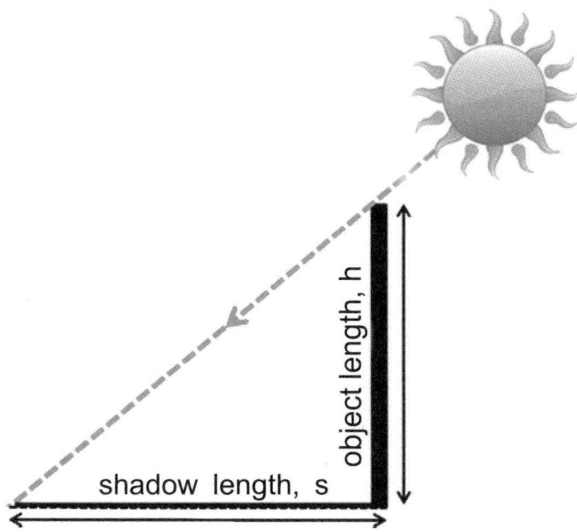

Fig. 2.4. Calculation of Air Mass using the shadow of an object of known height.

To standardize the measurement of solar radiation for solar energy applications, the American Society for Testing and Materials (ASTM) in 1982 developed "*Standard Tables of Reference Solar Spectral Irradiance at Air Mass 1.5*". The reference spectra were first established as separate standard, designated as E-891-82 and E-892-82 for direct normal and global tilt, respectively. This standard was again revised in 2003 as ASTM G173-03 by incorporating two spectra into a single spectrum. Further, an extension of this standard as ASTM G197-08 introduced corresponding to different tilt angles in 2008. The atmospheric and climatic conditions are identical in these both ASTM G173-03 and G197-08. These standards represent the solar radiation under reasonable cloudless atmospheric conditions favorable for the computerized simulation, comparative rating, or experimental testing of systems. The U.S. standard was adapted by the International Organization for Standardization (ISO) to become ISO 9845-1, 1992.

Fig. 2.5 shows the spectral irradiance or the radiation power spectrum for AM 0 and AM 1.5 solar radiation. The AM 0 spectrum is the solar radiation outside the atmosphere. It is approximately a black body radiation at 5,800 K. The integrated power density is 1.366 kW/m^2. The AM 1.5 spectrum shows a number of atmospheric effects.

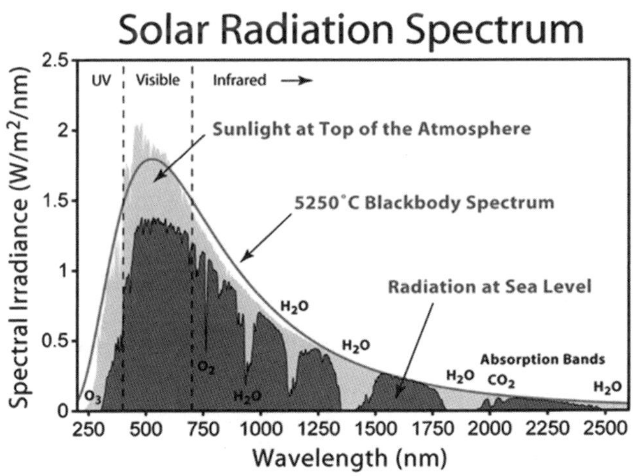

Fig. 2.5. Solar irradiance spectrum above atmosphere and at the earth's surface.

(Source: https://en.wikipedia.org/wiki/Solar_irradiance, Accessed on May 5, 2016 at 13.54 KST).

On the blue side, there is a broad-band reduction of power density due to Rayleigh scattering from molecules and dust particles.

The probability of Rayleigh scattering is proportional to the inverse fourth power of the wavelength of the radiation, and thus the short-wavelength radiation is reduced heavily. On the infrared side, water vapor contributes the most absorption followed by carbon dioxide.

Direct and diffuse radiation

After traveling in space, solar radiation enters the atmosphere at the space - atmosphere interface, where the ionization layer of the atmosphere ends. As discussed earlier, a part of solar radiation is absorbed by the atmosphere, clouds, and particles in the atmosphere, a part is reflected back into the space, and a remaining part is absorbed by the earth's surface. The earth's surface also reflects a part of radiation at different wavelengths due to the earth's surface temperature. About 50% of the total solar radiation remains in the atmosphere and earth's surface. The detailed percentages of each part of radiation can be seen in Fig. 2.2. The solar radiation falling on the earth is attenuated by the atmosphere and associated climate effects which change the amount of solar radiation received on the earth surface. As a combined effect of all the phenomenon such as reflection, absorption, refraction, and scattering, the solar radiation received at any point is both direct (or beam) and diffuse (or scattered). After the solar radiation enters the earth's atmosphere, it is partially scattered and partially absorbed. The partially absorbed radiation is called as direct radiation, where as the scattered radiation is called diffuse radiation. Again, a part of this diffuse radiation reflected back to space and a part reaches the earth's surface. Solar radiation reaches the earth's surface in two different ways as direct and diffuse, irradiations as in Fig. 2.6 and 2.7, respectively.

Direct solar radiation is that which travels in a straight line from the sun to the earth's surface (Fig. 2.6). Direct radiation also called as "Beam radiation" is the amount of solar radiation received at any place on the earth's surface directly from the sun without any disturbances. Practically, this radiation creates sharp shadows of the subjects. Direct solar radiation does not attenuate by dust, gas, and cloud or any other intermediate material. Direct radiation is the solar radiation received from the sun without scattered by the atmosphere.

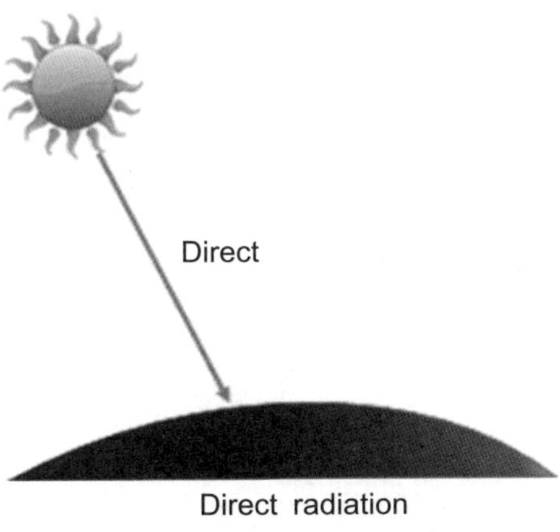

Fig. 2.6. Direct solar radiation paths.

The solar radiation undergoes wavelength- and direction-dependent adsorption and scattering by atmospheric gases, aerosols, and cloud droplets, while passing through the atmosphere. These scattered radiations reaching the earth's surface is called as diffuse radiation (Fig. 2.7). Diffuse radiation is first intercepted by the constituents of the air such as water vapor, CO_2, dust, aerosols, clouds, etc., and then released as scattered radiation in many directions. This diffuse radiation scattered in all direction does not give rise to sharp shadows even being close to earth's surface.

On a sunny clear day, the diffuse component from the Rayleigh and aerosol scattering is about 10 – 30% of the total incident radiation, whereas all the surface radiation is diffuse, when the solar radiation passes through a cloud. Diffuse radiation occurs when small particles and gas molecules diffuse part of the incoming solar radiation in random directions without any alteration in the wavelength of the electromagnetic energy.

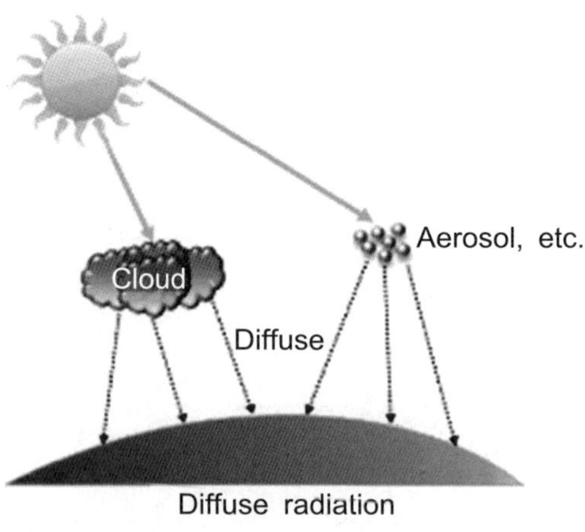

Fig. 2.7. Diffuse solar radiation paths.

Total solar radiation is the sum of the direct and the diffuse solar radiation on an earth's surface. The total radiation on a horizontal surface is most common measurements of solar radiation, which is hereafter referred to as global radiation on the earth's surface. The sun's total energy is composed of 7% ultraviolet (UV) radiation, 47% visible radiation, and 46% infrared radiation. UV radiation is significantly filtered out by the ozone layer in the upper atmosphere. The global radiation at the earth's surface consists of both short- and longwave radiation. Short-wave radiation may be absorbed by terrestrial bodies and cloud cover and re-emitted as long-wave radiation. The short-wave radiation reaching the surface of the earth may be direct, diffuse, or reflected (Fig. 2.6 and 2.7). Global radiation at a location is roughly proportional to direct solar radiation, and it varies with the nature of the receiving surface. The diffuse radiation varies slightly from slope to slope within a small area and the variations can be linked to slope gradient (14, 15). Generally, diffuse radiation contains less than 16% of the total solar irradiance at visible wavelengths in the green and red region (16), rising to 30% for blue.

Yearly solar insolation map

Solar insolation is a measure of solar radiation received on a given surface area in a given time. It is commonly expressed as average irradiance in watts per square meter (W/m^2) or kilowatt-hours per square meter per day ($kW \bullet h/(m^2 \bullet day)$) (or hours/day). It is commonly measured as kWh/(kWp • y) (kilowatt hours per year per kilowatt peak rating) in case of photovoltaics [17]. The solar insolation received at the earth's surface is controlled by the atmosphere, the angle of the sun, altitude, and geographic location. The actual solar radiation received at the surface is always less, because of the interaction of solar radiation with the atmosphere and the percentage in reduction depends on the location. The insolation onto an earth's surface is largest, when the surface directly faces the Sun. As the angle increases between the direction at a right angle to the surface and the direction of the rays of sunlight, the insolation is reduced in proportion to the cosine of the angle.

A commonly used representation is the annual solar insolation map of the world as well as solar insolation map of the South Korea as shown in the maps below. Fig. 2.8 is an average diurnal insolation map of the world; Fig. 2.9 is an average diurnal insolation map of the South Korea.

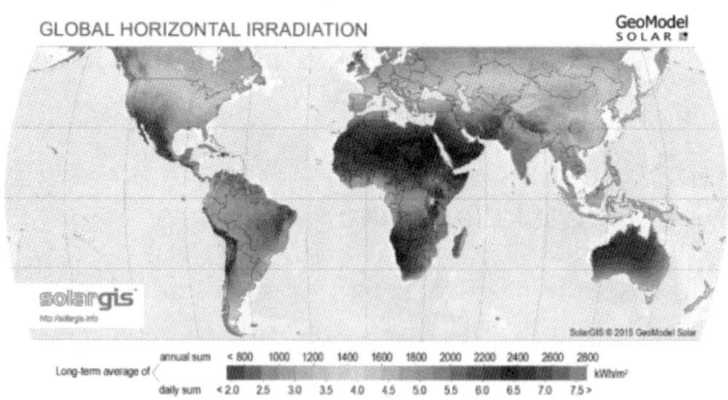

Fig. 2.8. Insolation map of the world.

(Source: http://solargis.info/doc/free-solar-radiation-maps-GHI#S, Accessed on May 5, 2016 at 15.25 KST)

Solar radiation per day on a surface of 1 m^2 in kilojoules averaged over a year. As shown, large areas in Northern Africa have the highest insolation.

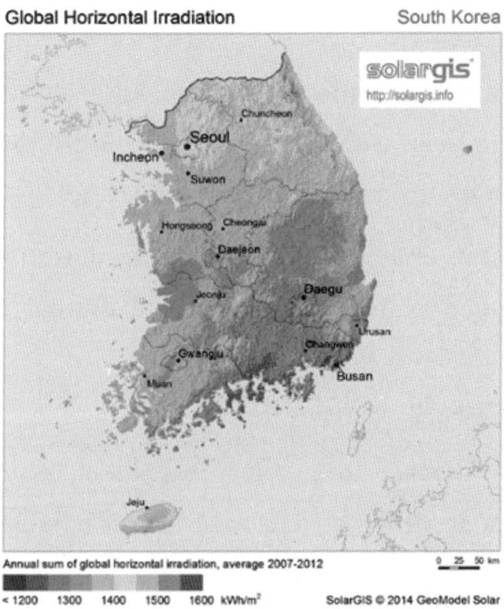

Fig. 2.9. Insolation map of the South Korea.

(Source: http://solargis.info/doc/free-solar-radiation-maps-GHI#S, Accessed on May 5, 2016 at 15.36 KST).

Solar radiation per day on a surface of 1 m^2 in kilojoules averaged over a year. As shown, east west states have the highest insolation. The northern part of the South Korea has rather weaker solar insolation than other areas.

There are two convections used [18]:

1. Annual radiation energy in kilowatt-hours per square meter. The standard solar radiation is defined as one sun, or 1 kW/m^2.

Therefore, the insolation is often expressed in hours per year. The number ranges from more than 2000 h/year (Sahara Desert, part of outback of Australia, part of South Africa) to less than 600 h/year (Greenland, northern parts of Siberia, Finland, and Canada).

2. Average diurnal radiation energy in kilowatt-hours per square meter over a year. Similarly, it is often expressed in h/day. The number ranges from more than 6 h/day to less than 2 h/day.

There is an obvious relation between those two conventions:

$$\text{Annual insolation} = 365.2422 \times \text{average diurnal insolation} \quad (2.2)$$

Measurement of solar radiation

Solar irradiance and solar insolation are the two most common methods which characterize the solar radiation. The solar irradiance is an instantaneous power density often measured in the units of kW/m^2 and it varies throughout the day from 0 kW/m^2 at night to a maximum of about 1 kW/m^2. On the other hand, the solar insolation is the total amount of solar energy received at a particular place during a particular time period, which is often measured in the units of $kWh/(m^2$ day). The solar insolation is quite different than the solar irradiance as the solar insolation is the average instantaneous solar irradiance measured over a given time period. Solar insolation data is commonly used for simple PV system design, whereas while solar radiance is used in more complicated PV system performance which calculates the system performance at each point in the day.

Fig. 2.10. A photograph of pyranometer. It complies with the "secondary standard" specifications within the latest ISO and WMO standards.

(Source: https://commons.wikimedia.org/wiki/File: Hukseflux_pyranometer_SR20-D1.png, Accessed on May 6, 2016 at 20.22 KST)

As discussed above, when solar radiation passes through the earth's atmosphere, some of it is absorbed or scattered by air molecules, water vapor, aerosols, and clouds. The solar radiation that received at the earth's surface directly is called direct normal irradiance (DNI), whereas the radiation that has been scattered out of the direct beam is called diffuse horizontal irradiance (DHI). The global horizontal irradiance falling on to the earth's surface consists of both these DHI from sky and DNI from the sun. These all are short-wave radiation and have a geometrical relationship as:

$$GHI = DHI + DNI * \cos\theta \tag{2.4}$$

where, θ is the solar zenith angle (vertically above the location $\theta = 0^o$ and horizontally $\theta = 90^o$).

GHI and DHI incident on an earth's surface are usually measured systematically by a national agency, i.e. by the national meteorological office in the country. GHI is measured by a broadband instrument such as pyranometer. A photograph of a typical pyranometer is represented in Fig. 2.10. It consists of a white disk for limiting the acceptance angle to 180° and two concentric hemispherical transparent covers made of glass. The two domes shield the thermopile detector from thermal convection, protect it against weather threat (rain, wind, and dust) and limit the spectral sensitivity of the instrument in the wavelength range 0.29 - 2.8 lm. A cartridge of silica gel inside the dome absorbs water vapor. The pyranometer for GHI may be conveniently mounted on top of the tracker. Such instruments have been used for a long time in meteorological and climatology networks and in the solar energy research community.

The pyrheliometer is a broadband instrument that used to measures a DNI. Consequently, this instrument should be permanently pointed toward the Sun. This instrument is a multi-junction thermopile type radiometer with a 5° field of view and a flat window placed at the bottom of a collimating tube provided with a quartz window to protect the instrument. The detector is coated with optical black paint (acting as a full absorber for solar energy in the wavelengths range 0.280 - 3 μm). Since, the pyrheliometer aperture angle is 5°, only radiation is received from the sun in a limited circumsolar region, but all diffuse radiation is excluded. Fig. 2.11 shows a photograph of pyrheliometer. This instrument must be mounted on a high accuracy automatic sun tracker in order to keep the radiometer pointed continuously at the centre of the sun.

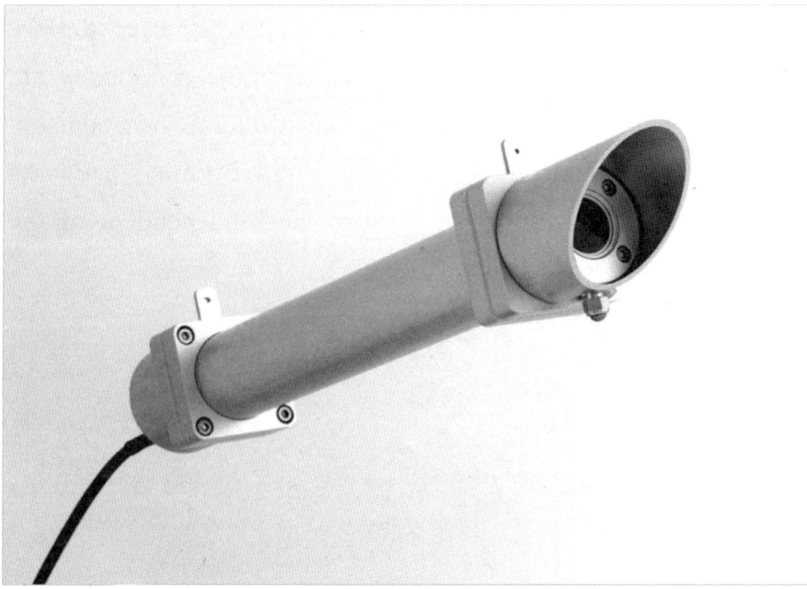

Fig. 2.11. A photograph of Hukseflux DR01 first class pyrheliometer.
(Source: https://en.wikipedia.org/wiki/Pyrheliometer#/media/File:Hukseflux_solarradiation_dr01_photo.jpg, Accessed on May 6, 2016 at 20.29 KST)

DHI can be measured by fitting a second pyranometer on top of the sun tracker and a shading assembly that moves with thetracker to always block the direct beam radiation from reachingthe pyranometer.

For the better solar radiation measurements, the World Meteorological Organization (WMO) and the International Standards Organization (ISO) have defined the different types of pyranometers and pyrheliometers, performance specifications, calibration methods, etc., The instruments which fulfill these criteria defined by the WMO and ISO, provide accurate measurements of the solar radiation under all weather conditions. Since the meteorological world and the scientific community are using the same types of instruments, the data can be compared with measurements from weather and climate station networks and satellites, across various locations, and independent of the type of solar energy system.

References

1. G. Boyle, "Renewable Energy", Second Edition, Oxford University Press, Oxford, UK, (2004).
2. P. D. Dunn, "Renewable energies: sources, conversion and application", Peregrinus, Cambridge, (1986).
3. J. F. Kreider and F. Kreith, "Solar energy handbook", McGraw-Hill, New York, (1981).
4. S. P. Langley, "Measurement of the direct effect of sun-spots on terrestrial climates", Mort. Notic. Roy. Astron. Soc., 37, 5-11 (1876).
5. S. P. Langley, "Researches on the solar heat and its absorption by the earth's atmosphere", A Report of the Mount Whitney Expedition, Prof. Pap. XV, Signal Serv., Washington, D.C., (1884).
6. S. P. Langley, "On a possible variation of the solar radiation and its probable effect on terrestrial temperatures", Astrophys. J., 19, 305-321 (1904).
7. G. Kopp and J. L. Lean, "A new, lower value of total solar irradiance: Evidence and climate significance", Geophys. Res. Lett., 38, L01706 (2011).
8. D. Archer, "Global Warming: Understanding the Forecast", ISBN 978-0-470-94341-0. (2012).
9. M. Stix, "The Sun, an Introduction", Second Edition, Springer, New York, (2002).
10. M. Naqi, "Encyclopaedia of Geomorphology", Anmol Publications

Pvt. Ltd.; (2006).

11. M. Iqbal, "An introduction to solar radiation", Academic, Toronto, 1986.
12. C. A. Gueymard, "The sun's total and spectral irradiance for solar energy applications and solar radiation models", Sol. Energy 76, 423-453 (2003).
13. C. Riordan, and R. Hulstron, "What is an air mass 1.5 spectrum? [solar cell performance calculations]", Photovoltaic Specialists Conference, 1990., Conference Record of the Twenty First IEEE. IEEE, 1990.
14. K. Y. Kondratyev, "Radiative heat exchange in the atmosphere", Pergamon, New York (1965).
15. L. D. Williams, R. G. Barry, J. T. Andrews, "Application of computed global radiation for areas of high relief", J. Appl. Meteorol. 11, 526-533 (1972).
16. R. Dubayah, "Estimating net solar radiation using Landsat Thematic Mapper and digital elevation data", Water Resour. Res. 28, 2469-2484 (1992).
17. N. Robinson, "Solar Radiation", Elsevier, Amsterdam, 1976.
18. C. J. Chen, "Physics of Solar Energy", John Wiley & Sons, Inc. Hoboken, New Jersey, 2011.

Chapter 3
Basics of semiconductor physics

3.1. Binding forces in solids

Each atom is made of electrons, protons, and neutrons. Electron and protons carry negative and positive charges of the same magnitude, 1.6×10^{-19} Coulombs. Fig. 3.1 shows a typical atomic structure containing electrons, protons, and neutrons.

When we compare the mass of the electron it is negligible with respect to those of the proton and the neutron, which form the *nucleus* of the atom.

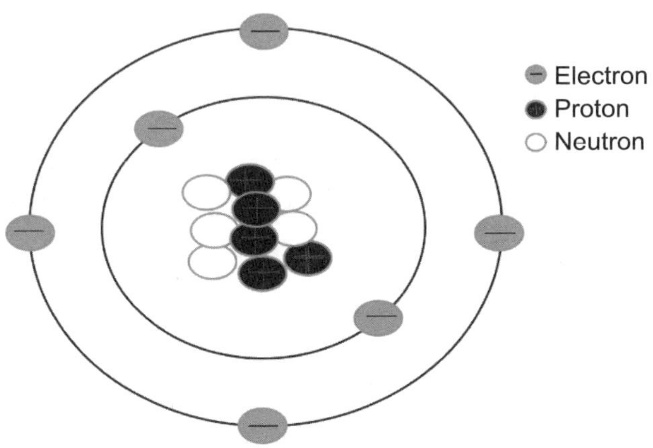

Fig. 3.1. A typical carbon atom which consists its constituent elements.

The mass of the electron is negligible with respect to those of the proton and the neutron, which form the nucleus of the atom. The unit of mass is an atomic mass unit (amu) = 1.66×10^{-27} kg, and equals 1/12 the mass of a carbon atom. The Carbon nucleus has Z=6, and A=6, where Z is the number of protons, and A the number of neutrons. Neutrons and protons have very similar masses, roughly equal to 1 amu. A neutral atom has the same number of electrons and protons, Z.

A mole is the amount of matter that has a mass in grams equal to the atomic mass in amu of the atoms. Thus, a mole of carbon has a mass of 12 grams. The number of atoms in a mole is called the Avogadro number, $N_{av} = 6.023 \times 10^{23}$. Note that $N_{av} =$ 1 gram/1 amu.

Calculating n, the number of atoms per cm^3 in a piece of material of density δ (g/cm^3).

$$n = N_{av} \times \delta / M$$

where M is the atomic mass in amu (grams per mol). Thus, for graphite (carbon) with a density $\delta = 1.8\ g/cm^3$, M =12, we get 6×10^{23} atoms/mol $\times$ 1.8 g/cm^3 / 12 g/mol) = 9×10^{22} C/cm^3.

For a molecular solid like ice, one uses the molecular mass, $M(H_2O) = 18$. With a density of 1 g/cm^3, one obtains n = 3.3×10^{22} H_2O/cm^3. Note that since the water molecule contains 3 atoms, this is equivalent to 9.9×10^{22} atoms/cm^3.

Most solids have atomic densities around 6×10^{22} atoms/cm^3. The cube root of that number gives the number of atoms per centimeter, about 39 million. The mean distance between atoms is the inverse of that, or 0.25 nm. This is an important number that gives the scale of atomic structures in solids.

3.1.1. Electrons in atoms

There are repulsive forces between the electrons in atoms and attractive between electrons and protons. The neutrons play no significant role.

There is a cloud of electrons around the neutrons of radius of 0.05 to 2 nm. The motion of electrons is not circular but it is in 'fuzzy' orbits. It means that, it is the probability of finding an electron at some distance from the nucleus. Based on quantum mechanics, only certain orbits are allowed. These orbits are identified by a principal quantum number n, which can be related to the size, n = 0 is the smallest; n = 1, 2. are larger. The angular momentum 'l' is quantized, and so is the projection in a specific direction m. The structure of the atom is determined by the Pauli exclusion principle, which states 'only two electrons can be placed in an orbit with a given n, l, m – one for each spin'.

3.1.2. Bonding forces and energies

There exist simple Coulomb forces between electrons and nuclei. It is attractive between electrons and nuclei and repulsive between each electron-electron and each nuclei-nucleus. The sum of all the individual forces defines the force between the atoms. There is nucleus in the center of an atom and electrons are located outside an atom.

If two atoms come very close, the force between them is always repulsive, this is due to the electrons are outside an atom and the nuclei repel each other. When both atoms are ions of the same charge then the forces between atoms is always attractive at large inter-nuclear distances r. The reason is the force is repulsive at small r, and attractive at small r, there is a distance at which the force is zero. This distance is the equilibrium distance where atoms prefer to stay (Fig. 3.2).

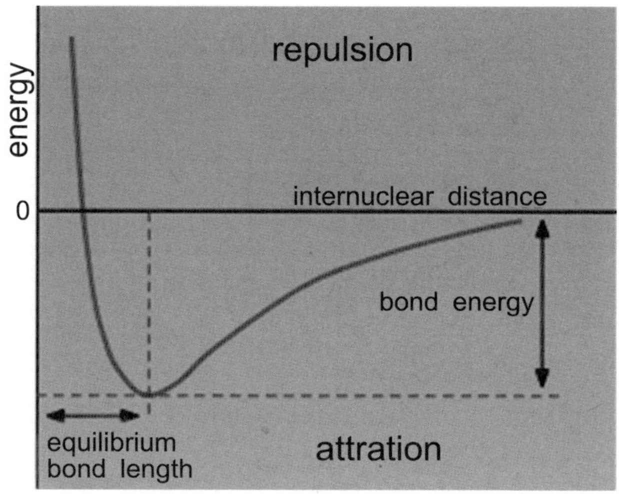

Fig. 3.2. Bonding force and energies between atoms.

(Source:http://medsocnet.ncsa.illinois.edu/MSSW/tutinf/Webmo/introduction/intro.html)

The interaction energy between the atoms is called as the potential energy. It is negative when the atoms are bound and is positive when atoms can move away from each other. The interaction energy is the sum of the force over the separation distance, so these two quantities are interdependent. The interaction energy is a minimum at the equilibrium position. This value of the energy is called the bond energy, and is the energy needed to separate completely to infinity. When the bond energy is the strongest it is harder to move the atoms. Therefore it is hardest to melt the solid or to evaporate its atoms.

3.1.3. Primary interatomic bonds

3.1.3.1. Ionic bonding

It is the kind on bond where one of the atoms is negative having extra electron and the other is positive which has lost an electron. Hence, there is a strong, direct Coulomb attraction. An example is NaCl (Fig. 3.3). In the NaCl molecule, there are more electrons around Cl, forming Cl^- and less around Na, forming Na^+. Ionic bonds are the strongest bonds. In real solids, ionic bonding is usually combined with covalent bonding. In this case, the fractional ionic bonding is defined as % ionic = 100 × [1 − exp(− 0.25 $(X_A - X_B)^2$], where X_A and X_B are the electronegativities of the two atoms, A and B, forming the molecule.

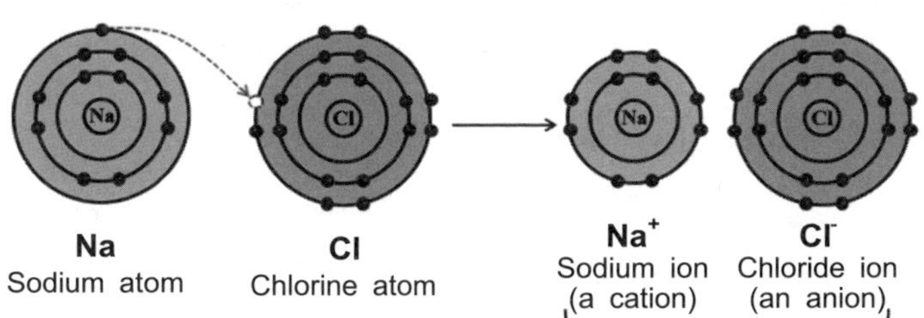

Fig. 3.3. Ionic bonding in NaCl crystal.

3.1.3.2. Covalent bonding

In case of covalent bonding, electrons are shared between the molecules, in order to saturate the valency. The simplest example is the H_2 molecule (Fig. 3.4 (a)), in which the electrons spend more time in between the nuclei than outside, thus producing bonding.

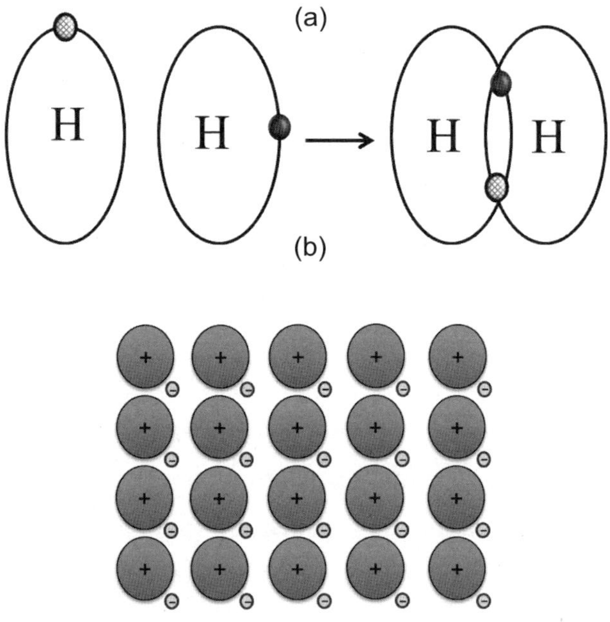

Fig. 3.4. (a) Covalent bonding and (b) metallic bonding in H_2 molecule.

3.1.3.3. Metallic bonding

In case of metals, the atoms are ionized, losing some electrons from the valence band. Those electrons form an electron sea, which binds the charged nuclei in place, in a similar way that the electrons in between the H atoms in the H_2 molecule bind the protons (Fig. 3.4 (b)).

3.1.4. Secondary bonding (Van der Waals)

3.1.4.1. Fluctuating induced dipole bonds

When the electrons are on the one side or the other side of the atom, a dipole is formed: the + nucleus at the center, and the electron outside. As the electron moves, the dipole fluctuates. This fluctuation in atom A produces a fluctuating electric field which is felt by the electrons of an adjacent atom, B. Hence, Atom B polarizes so that its outer electrons are on the side of the atom closest to the + side (or opposite to the – side) of the dipole in A. This bond is called van der Waals bonding (Fig. 3.5).

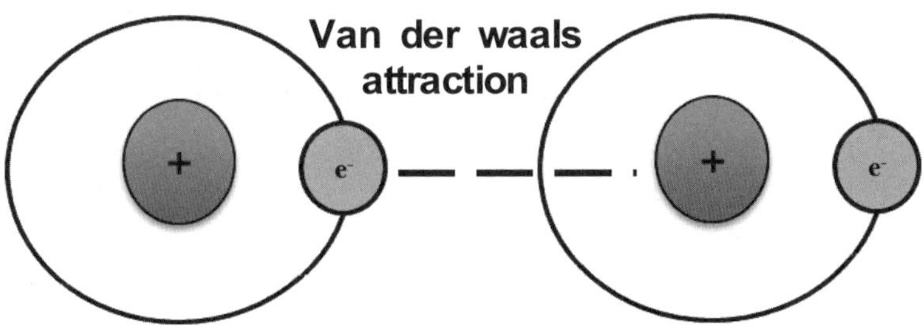

Fig. 3.5. van der Waals bonding.

3.1.4.2. Polar molecule-induced dipole bonds

When a polar molecule likes H_2O (H atoms are partially +, O is partially –), will induce a dipole in a nearby atom, it leads to bonding. Such bonding is referred as polar molecule-induced dipole bonds as shown in Fig. 3.6 (a).

3.1.4.3. Permanent dipole bonds

This is the case of the hydrogen bond in ice. The H end of the molecule is positively charged and can bond to the negative side of another dipolar molecule, like the O side of the H_2O dipole (Fig. 3.6 (b)).

(a)

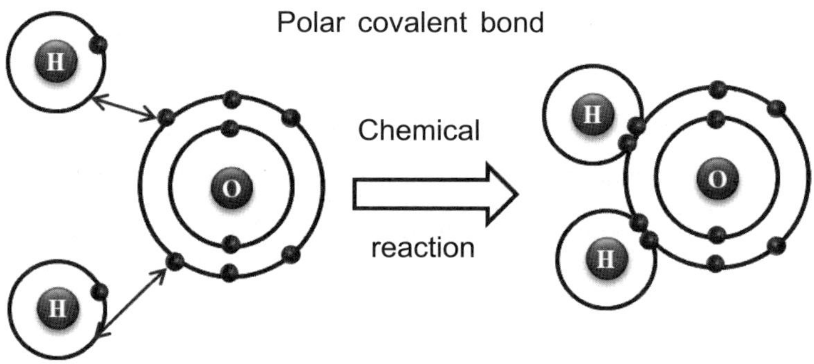

(b)

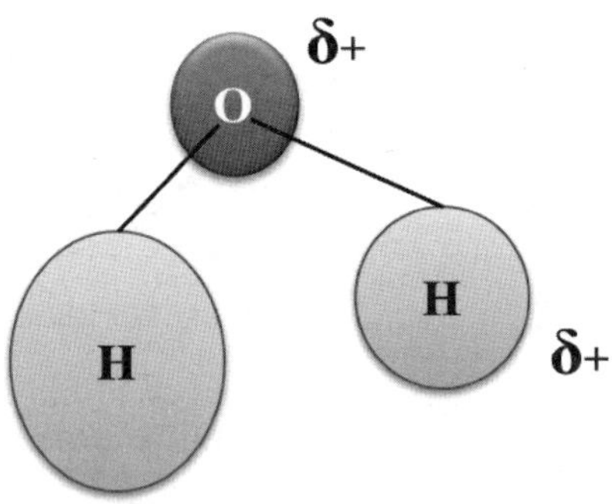

Fig. 3.6. (a) Polar Molecule-Induced Dipole Bonds in H_2O
(b) Permanent Dipole Bonds.

3.1.4.4. Molecules

When molecules form a closed shell due to covalent bonding (like H_2, N_2) then the interaction between molecules is weak, of the van der Waals type. Therefore, molecular solids usually have very low melting points.

3.2. Classification of solids

3.2.1. Conductors

Metals are good electrical conductors. In metals the valence band is partially empty and not completely full of electrons. The valence band of metals contains empty states above and immediately adjacent to the Fermi level into which electrons can be excited (Fig. 3.7). These states require only miniscule energy

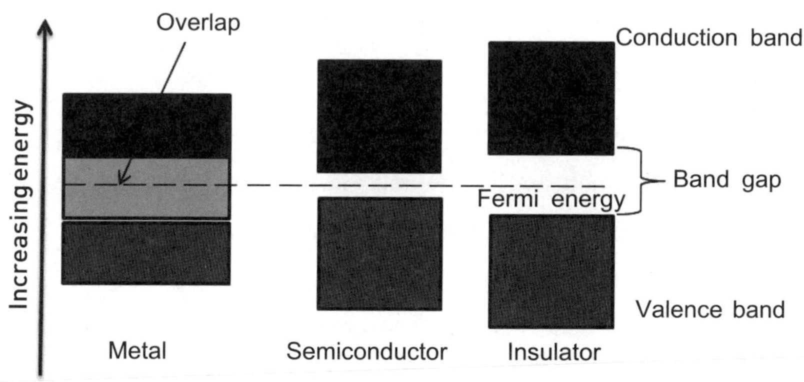

Fig. 3.7. Energy bands of metal, semiconductor and insulator.

excitation from the filled states at the Fermi level. Most of the thermal, electrical and optical properties of metals depend on this low excitation energy.

There are two general classes of energy band diagrams for metals: In case of monovalent alkali metals such as sodium, the 3s energy band includes twice as many levels as there are Na atoms; but, only half of these levels are filled.

In case of divalent alkaline earth metals such as magnesium the 3s and 3p energy bands overlap; this offers many empty states where electrons can be easily excited.

Hence we can say all metals are conductors and that all conductors are metals. But the first generalization is correct and the second is not. Consider for example the transition metal oxides such as CrO_2. This solid has a partially filled and extended energy band involving overlap of d and f levels resulting in the ability for conduction. Thus there are both metallic and ceramic conductors.

3.2.2. Semiconductors

The most general semiconducting materials are the covalently bonded solids such as Si, Ge, and GaAs. Fig. 3.8 shows periodic table of elements consisting semiconductor materials which are shown in light blue color. There are variety of other band gap solids with their band gap equal to or greater than 2.5 eV including ceramics such as SiC and organic materials such as anthracene (coal). Let's consider the $1s^2 2s^2 2p^6 3s^2 3p^2$ electronic structure of Si. Si is covalently bonded with each atom having four sp^3 hybridized bonds. Therefore, each atom contributes four levels to the valence and conduction bands, at 0 K, the valence band is completely filled, the conduction band is completely empty, and the Fermi energy level falls in the gap. Because of small energy gap in semiconductors, an increase in temperature permits electron to move to the conduction band. However, the gap allows only a certain number of electrons to take part to the conduction process.

Compared to typical metals a pure semiconductor like Si is about 10^{10} times a poorer conductor at room temperature. The electronic properties of the semiconductors are driven by the controlled addition of impurities called as doping. When atoms with less than 4 sp^3 electrons are added as impurities like Ga and Ge

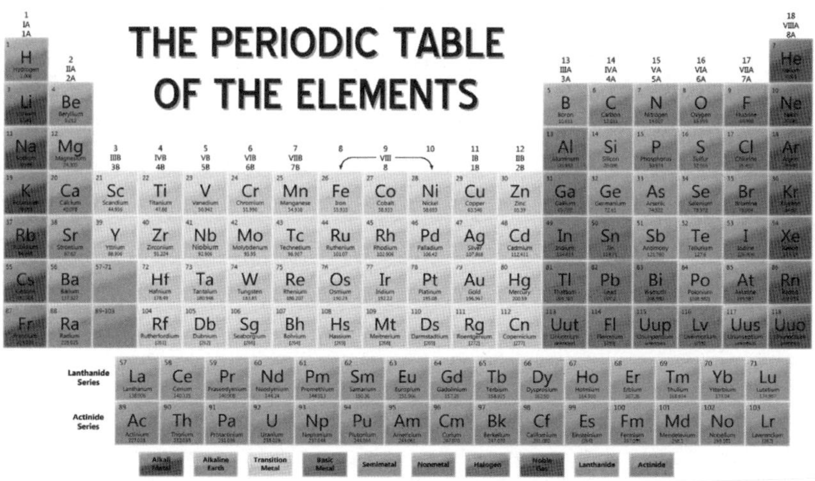

Fig. 3.8. Periodic table of elements with semiconductors shown in light blue shed and marked as semimetals.

(Source: http://sciencenotes.org/printable-periodic-table/)

then there appears additional sates in the forbidden gap near the upper edge of the valence band. While if atoms with more than 4 sp^3 electrons are added as impurities like As in Ge, then there appears additional states in the forbidden gap near the bottom edge of the conduction band.

3.2.3. Insulators

Generally most pure metal oxides like Al_2O_3, the silicate ceramics and the common organic polymers are considered as insulators. Since the Fermi energy falls inside the energy gap in the case of semiconductors; however, the energy gap for insulators is so large that in most cases it cannot easily be overcome by thermal energy or the application of an electric field.

3.2.4. Types of semiconductors

Semiconductors are mainly classified into two categories: Intrinsic and Extrinsic.

3.3. Intrinsic semiconductors

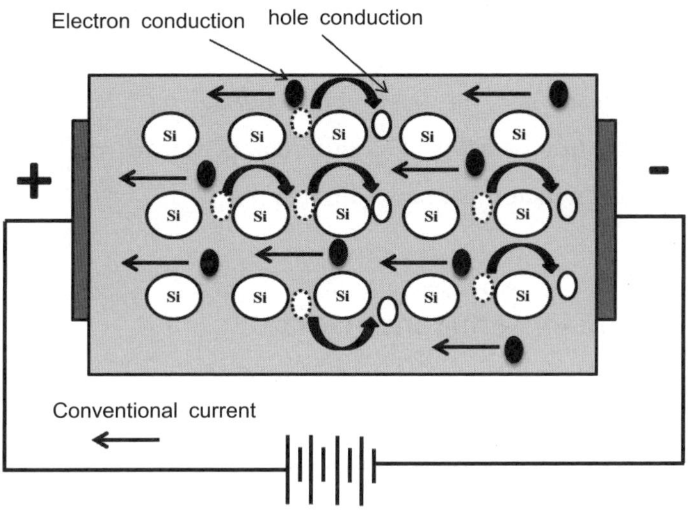

Fig. 3.9. Intrinsic semiconductors.

A material which is chemically very pure and exhibits poor conductivity is called as intrinsic semiconductor (Fig. 3.9). It has even numbers of negative carriers (electrons) and positive carriers (holes). A silicon crystal is different from an insulator because at any temperature above absolute zero temperature, there is a finite probability that an electron in the lattice will be knocked loose from its position, leaving behind an electron deficiency called a "hole".

If a voltage is applied, then both the electron and the hole can contribute to a small current flow.

Band theory of solids explains the conductivity of a semiconductor. The band model of a semiconductor proposes that at ordinary temperatures there is a finite possibility that electrons can reach the conduction band and contribute to electrical conduction.

The meaning of intrinsic here refers to pure silicon which distinguishes from surprisingly different properties of n-type or p-type semiconductors because of doping.

3.4. Extrinsic semiconductors

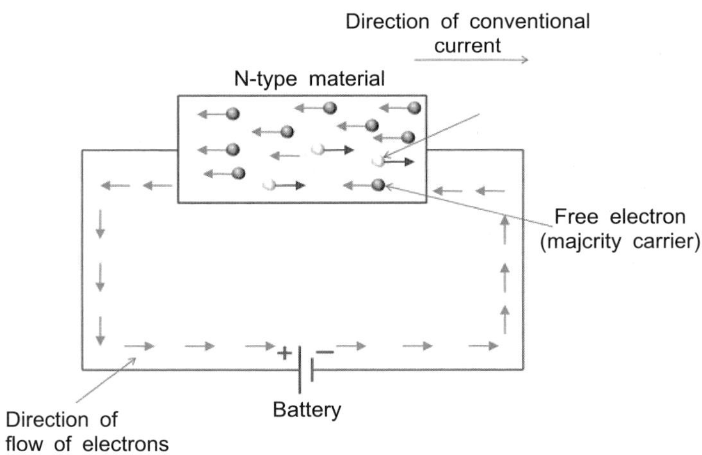

Fig. 3.10. Extrinsic semiconductors.

An extrinsic semiconductor is an enriched intrinsic semiconductor with a small amount of impurities added by a process, known as doping; this changes the electrical properties of the semiconductor and advances its conductivity (Fig. 3.10). Presenting impurities into the semiconductor materials (doping process) we are able to regulate their conductivity.

Doping process results two groups of semiconductors: those are the negative charge carrier (n-type) and the positive charge carrier (p-type). There exists either elements or compounds semiconductors. Silicon and Germanium are the most common elemental semiconductors. Compound Semiconductors include InSb, InAs, GaP, GaSb, GaAs, SiC, GaN. Si and Ge both have a crystalline structure called the diamond lattice. That is, each atom has its four nearest neighbors at the corners of a regular tetrahedron with the atom itself being at the center. In addition to the pure element semiconductors, many alloys and compounds are semiconductors. The benefit of compound semiconductor is that they provide the device engineer with a variety of energy gaps and mobilities, so that materials are available with properties that meet particular requirements. Some of these semiconductors are therefore referred wide band gap semiconductors.

3.4.1. The doping of semiconductors

In order to produce n-type and p-type semiconductors one need to add small percentage of foreign atoms in the crystal lattice of semiconductors like silicon or germanium. Doping of semiconductors leads to changes in their electrical properties (Fig. 3.11). We can do doping by two kinds viz., using Pentavalent and trivalent impurities. Pentavalent impurities contain 5 valence electrons and they produce n-type semiconductors by contributing extra electrons. While trivalent impurities contain 3 valence electrons which produce p-type semiconductors by producing a "hole" or electron deficiency.

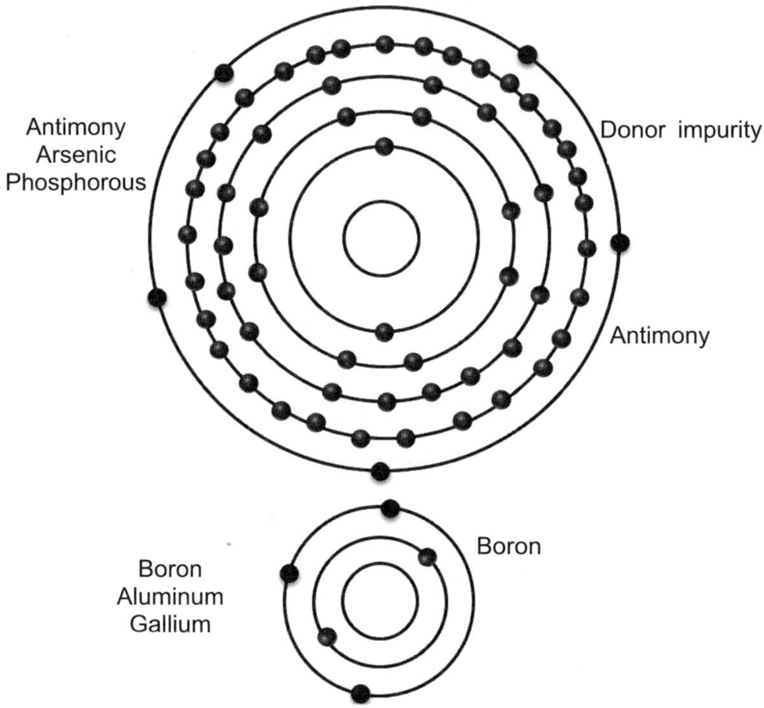

Fig. 3.11. Doping in semiconductors.

3.4.2. N-Type semiconductor

The conductivity of the intrinsic semiconductors can greatly be enhanced by the addition of pentavalent impurities such as antimony, arsenic or phosphorous which contributes free electrons (Fig. 3.12 (a)). It is possible to add phosphorous by diffusion of phosphine gas (PH_3).

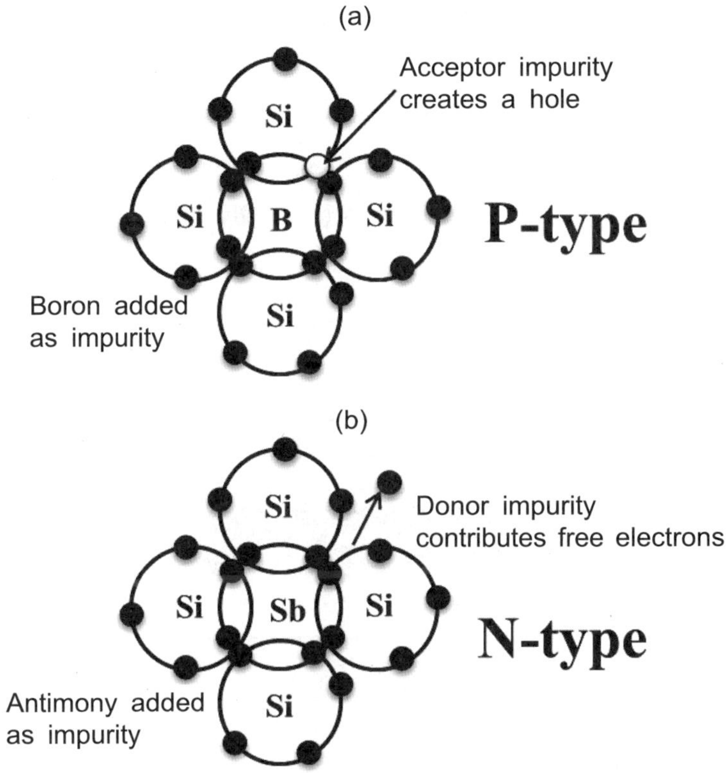

Fig. 3.12. (a) P-type semiconductor (b) N-type semiconductor.

3.4.3. P-Type semiconductor

In intrinsic semiconductors deficiencies of valence electrons can be create by adding trivalent impurities such as boron, aluminum or gallium (Fig. 3.12 (b)). Such deficiencies of electrons are called as “holes”. Boron into the silicon material is diffused using B_2H_6 diborane gas.

3.5 Carrier concentration

For an intrinsic semiconductor, the concentration of electrons in the conduction band is equal to the concentration of holes in the valence band (Fig. 3.12 (c)). We may denote n_i and p_i as the electron and hole concentrations, respectively in the intrinsic semiconductor. These parameters are usually referred to as the intrinsic electron concentration and the intrinsic hole concentration.

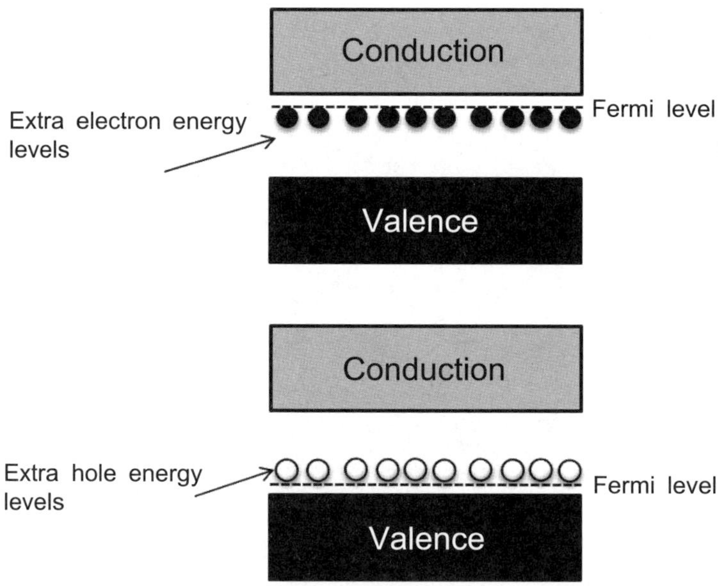

Fig. 3.12. (c) Fermi level for N-type semiconductor
(d) Fermi level for P-type semiconductor.

However $n_i = p_i$ so normally we use only the parameter n_i which refer to either intrinsic electron or hole concentration. The Fermi energy level for the intrinsic semiconductor is called the intrinsic Fermi energy, or $E_F + E_{Fi}$.

If we can write

$$n_o = n_i = N_c e^{-(E_c k_T - E_{Fi})} \tag{3.1}$$

and

$$p_0 = p_i = n_i = N_v e^{-(E_{Fi} k_T - E_v)} \tag{3.2}$$

If we take the product of Equations (3.1) and (3.2), we obtain

$$n_i^2 = N_c N_v e^{-(E_c k_T - E_{Fi})} e^{-(E_{Fi} k_T - E_v)} \tag{3.3}$$

or

$$n_i^2 = N_c N_v e^{-(E_c k_T - E_v)} = N_c N_v e^{-(k_T - E_g)} \tag{3.4}$$

where E_g is the bandgap energy.

For a given semiconductor material at a constant temperature, the value of n_i is a constant, and independentof the Fermi energy. The intrinsic carrier concentration for silicon at T = 300 K may be calculated by using the effective density of states function values. The value of n_i calculated from Equation (3.4) for E_g = 1.12 eV is n_i = 6.95 × 10^{19} cm^{-3}. The commonly accepted value of n_i for silicon at T = 300 K is approximately 1.5 × 10^{10} cm^{-3}. This discrepancy may arise from several sources.

First, the values of the effective masses are determined at a low temperature where the cyclotron resonance experiments are performed. Since the effective mass is an experimentally determined parameter, and since the effective mass is a measure of how well a particle moves in a crystal, this parameter may be a slight function of temperature. Next, the density of states function for a semiconductor was obtained by generalizing the model of an electron in a three-dimensional infinite potential well. This theoretical function may also not agree exactly with experiment. However, the difference between the theoretical value and the experimental value of n_i is approximately a factor of 2, which, in many cases, is not significant.

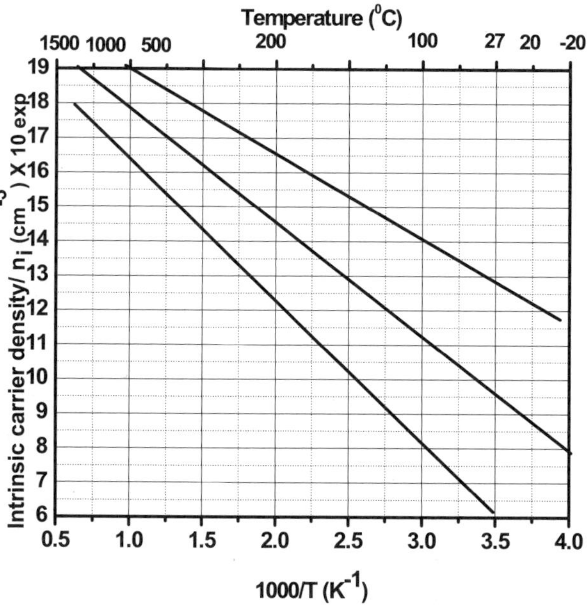

Fig. 3.13 Plot of n_i for Si, GaAs and Ge.

Table 1 lists the commonly accepted values of n_i for silicon, gallium arsenide, and germanium at T = 300 K. The intrinsic carrier concentration is a very strong function of temperature.

Table 1. commonly accepted values of n_i at T = 300 K

Semiconductor	Carrier concentration (n_i) at 300 K
Silicon	1.5×10^{10} cm^{-3}
Gallium Arsenide	1.8×10^{6} cm^{-3}
Germanium	2.4×10^{13} cm^{-3}

Fig. 3.13 is a plot of n_i from Equation (3.4) for silicon, gallium arsenide, and germanium as a function of temperature. As seen in the figure, the value of n_i for these semiconductors may easily vary over several orders of magnitude as the temperature changes over a reasonable range.

3.6. Mobility and conductivity

In a semiconductor the carrier mobility is directly linked to the conductivity and resistivity. Here we examine the doping dependence of the mobility and the corresponding doping dependence of the conductivity and resistivity.

The mobility of electrons and holes in silicon at room temperature is shown in Fig. 3.14.

Doping dependence is similar in case of the electron mobility and hole mobility: The mobility is almost constant in low doping concentrations, and is primarily limited by phonon scattering. Ionized impurity scattering with the ionized doping atoms leads to higher doping concentrations, hence the mobility decreases. The mobility is dependent to type of dopant.

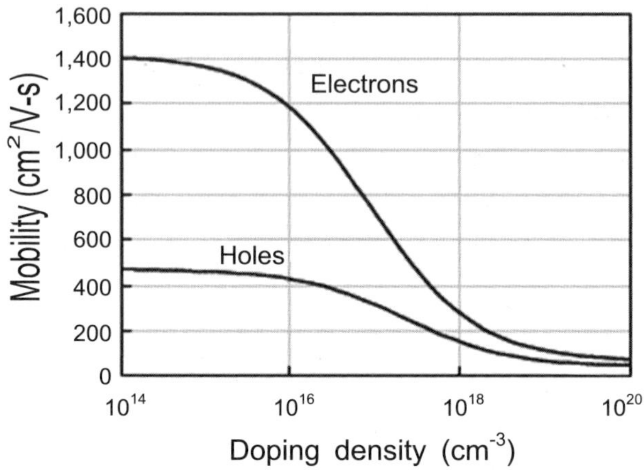

Fig. 3.14. The mobility of electrons and holes at RT.

(Source: http://ecee.colorado.edu/~bart/book/book/chapter2/ch2_7.htm)

The mobility is associated to the sum of the donor and acceptor densities or total number of ionized impurities. The free carrier density, is to first order related to the difference between the donor and acceptor concentration.

The minority carrier mobility also depends on the total impurity density. The minority-carrier mobility can be approximated by the majority-carrier mobility in a material with the same number of impurities.

The mobility at a particular doping density is obtained from the following empiric expression:

$$\mu = \mu_{\min} + \frac{\mu_{\max} - \mu_{\min}}{1 + \left(\frac{N}{N_r}\right)^{\alpha}}$$

Where $\mu_{\min}$, $\mu_{\max}$, α and N_r are fit parameters. These parameters for arsenic, phosphorous and boron doped silicon are provided in Table 2.

Table 2. Parameters for calculation of the mobility as a function of the doping density.

	Arsenic	Phosphorous	Boron
$\mu_{\min}$ (cm^2/V-s)	52.2	68.5	44.9
$\mu_{\max}$ (cm^2/V-s)	1417	1414	470.5
N_r (cm^{-3})	9.68×10^{16}	9.20×10^{16}	2.23×10^{17}
α	0.68	0.711	0.719

Table 3. Mobility in silicon for different dopants and doping densities

N	Arsenic μ_n (cm^2/V-s)	Phosphorous μ_n (cm^2/V-s)	Boron μ_p (cm^2/V-s)
10^{17} cm^{-3}	52.2	68.5	44.9
10^{16} cm^{-3}	1417	1414	470.5
10^{17} cm^{-3}	9.68×10^{16}	9.20×10^{16}	2.23×10^{17}
10^{18} cm^{-3}	0.68	0.711	0.719
10^{19} cm^{-3}	108	115	71

The resulting mobilities in units of cm^2/V.s are listed for different doping densities in Table 3.

3.6.1. Conductivity and resistivity

The conductivity of a material is defined as the current density divided by the applied electric field. Since the current density equals the product of the charge of the mobile carriers, their density and velocity, it can be expressed as a function of the electric field using the mobility. To include the contribution of electrons as well as holes to the conductivity, we add the current density due to holes to that of the electrons, or:

$$J = qnv_e + qnv_e = q(n\mu_n + p\mu_p)\epsilon \tag{3.5}$$

The conductivity due to electrons and holes is then obtained from:

$$\sigma \overset{\triangle}{=} \frac{J}{\epsilon} = q(n\mu_n + p\mu_p) \tag{3.6}$$

The resistivity is defined as the inverse of the conductivity, namely:

$$\rho = \frac{1}{\sigma} = \frac{1}{q(\mu_n n + \mu_p p)} \tag{3.7}$$

The resulting resistivity as calculated with the expression above is shown in Fig. 3.15.

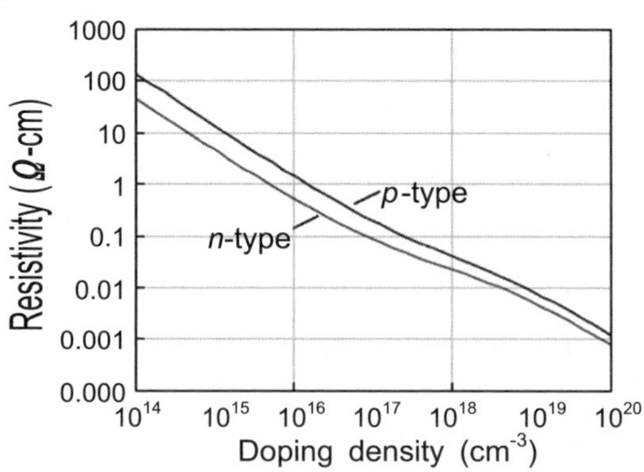

Fig. 3.15. Resistivity of n-type and p-type silicon versus doping density.

(Source:http://ecee.colorado.edu/~bart/book/book/chapter2/ch2_7.htm)

3.6.2. Sheet resistance

In order to characterize both wafers and thin doped layers the sheet resistance is used, since it is typically simpler to measure the sheet resistance rather than the resistivity of the material. The sheet resistance of a uniformly-doped layer with resistivity, R, and thickness, t, is given by their ratio:

$$R_s = \frac{\rho}{t} \tag{3.8}$$

The unit of sheet resistance is Ohms but usually referred as Ohms per square. This nomenclature comes in convenient when the resistance of a rectangular piece of material with length, L, and width W must be obtained. It equals the product of the sheet resistance and the number of squares or:

$$R = R_s \frac{L}{W} \tag{3.9}$$

Where the number of squares equals the length divided by the width. Fig. 3.16 provide, as an example, the sheet resistance of a 14 mil thick silicon wafer for both n-type and p-type silicon.

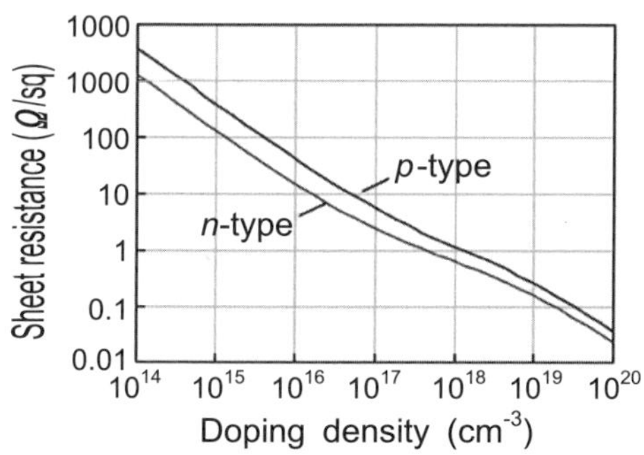

Fig. 3.16. Sheet resistance of a 14 mil thick n-type and p-type silicon wafer versus doping density.

(Source: http://ecee.colorado.edu/~bart/book/book/chapter2/ch2_7.htm)

A piece of silicon doped with arsenic ($N_d = 10^{17}$ cm^{-3}) is 100 mm long, 10 mm wide and 1 mm thick. Calculate the resistance of this sample when contacted one each end.

The resistivity of the silicon equals:

$$\rho = \frac{1}{qu\mu_n} = \frac{1}{1.6 \times 10^{-19} \times 10^{-17} \times 727} = 0.086\ \Omega\text{cm}$$

The resistance then equals:

$$R = \rho \frac{1}{W_t} = 0.086 \frac{100 \times 10^{-4}}{10 \times 10^{-4} \times 10^{-4}} = 8.6 k\Omega$$

An alternate approach is to first calculate the sheet resistance, R_s :

$$R_s = \frac{\rho}{t} = \frac{0.086}{10^{-4}} = 860\ \Omega/\text{square}$$

From which one then obtains the resistance:

$$R = R_S \frac{L}{W} = 860 \times \frac{100 \times 10^{-4}}{10 \times 10^{-4}} = 8.6 k\Omega$$

Next, we generalize the concept of the sheet resistance to a semiconductor layer with non-uniform doping as illustrated with Fig. 3.17.

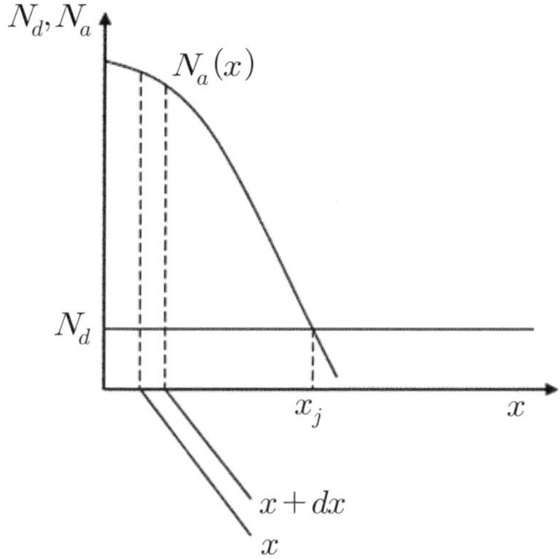

Fig. 3.17. Non-uniform doping distribution.

(Source: http://ecee.colorado.edu/~bart/book/book/chapter2/ch2_7.htm)

The doping profile is representative of a diffused and/or ion-implanted p-type layer in a uniformly doped n-type substrate. The metallurgical junction depth x_j occurs when the two doping densities are equal. The sheet resistance of the section with width, dx, is given by:

$$dR_S = \frac{\rho}{dx} = \frac{1}{q\mu_p(x)p(x)dx} \tag{3.10}$$

Where both the mobility, $\mu_p(x)$and hole density $p(x)$ are dependent on position. The sheet resistance of the whole layer is the parallel resistance of all slices with width dx between zero and x_j resulting in:

$$R_s = \frac{1}{\int_0^{x_t} \mu_p(x)p(x)dx} \cong \frac{1}{q\overline{\mu_p}\int_0^{x_t} N_a(x)dx} \tag{3.11}$$

Where $\overline{\mu_p}$ is the average mobility in the doped layer.

3.7. Drift and diffusion

Drift and diffusion currents: -

In a semiconductor material the flow of charge current is of two types namely drift & diffusion. The resultant current that flows through a (p-n junction diode) semiconductor material has two components

(i) Drift current

(ii) Diffusion current

(i) Drift current: -

When an electric field is applied across the semiconductor material, the charge carriers attain a certain drift velocity V_d, which is equal to the product of the mobility of the charge carriers and the applied electric field intensity E;

Drift velocity V_d = mobility of the charge carriers × applied Electric field intensity.

Holes migrate near the negative terminal of the battery and electrons migrate near the positive terminal of the battery. This collective effect of movement of the charge carriers establishes a current known as "the drift current"

Thus the drift current is defined as the flow of electric current due to the motion of the charge carriers when an external electric field is imposed. Drift current due to the charge carriers such as free electrons and holes are the current passing through a square centimeter perpendicular to the direction of flow.

Drift current density J_n, due to free electrons is given by

$$J_n = q_n \mu_n \text{ E A/cm}^2 \tag{3.12}$$

Drift current density J_P, due to holes is given by

$$J_p = q_p \mu_p \text{ E A/cm}^2 \tag{3.13}$$

Where, n-Number of free electrons per cubic centimeter.

P-Number of holes per cubic centimeter

μ_n – Mobility of electrons in cm^2/Vs

μ_p – Mobility of holes in cm^2/ Vs

E – Applied Electric filed Intensity in V/cm

q – Charge of an electron = 1.6×10^{-19} coulomb.

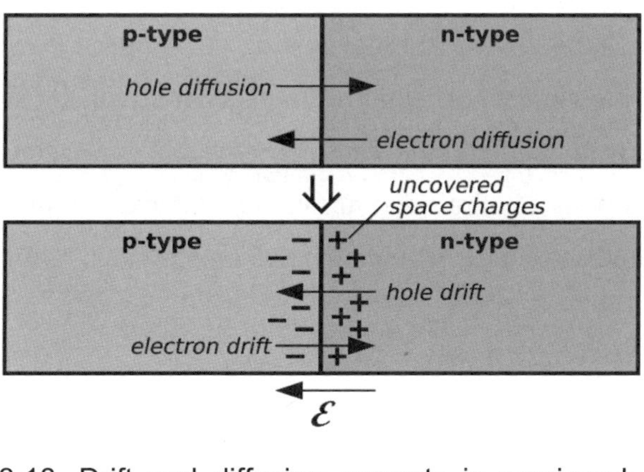

Fig. 3.18. Drift and diffusion currents in semiconductors.

(ii) Diffusion current: -

If there is absence of the applied electric field an electric current can flow in a semiconductor when there exists a concentration gradient. There is a concentration gradient when the number of either elements or holes is greater in one region of a semiconductor as compared to the rest of the region. In a semiconductor material the charge carriers have the tendency to move from the region of higher concentration to that of lower concentration of the same type of charge carriers. Thus the movement of charge carriers takes place resulting in a current called diffusion current.

3.8. Optical absorption

The band gap is defined as the minimum energy difference between the top of the valence band and the bottom of the conduction band. However, the top of the valence band and the bottom of the conduction band are not generally at the same value of the electron momentum. The energy band of a semiconductor material consists of three distinct energy bands, conduction band, valence band and the forbidden band as shown in the Fig. 3.19 below.

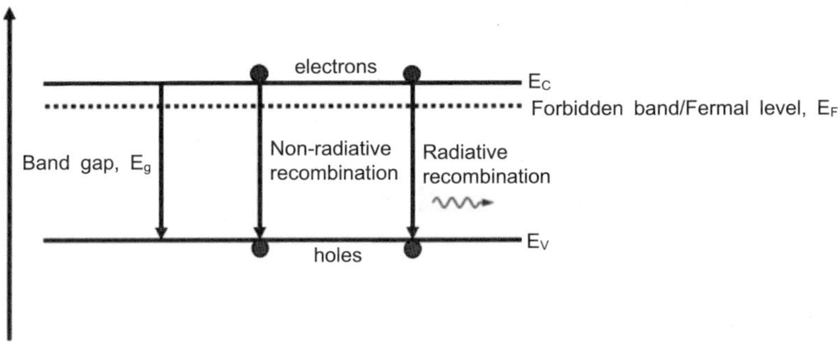

Fig. 3.19. Energy band of a semiconductor.

Therefore based on this fact that is whether they possess same value of momentum or different band gap is categorized into following two types-Direct band gap semiconductor and Indirect band gap semiconductor.

As shown in the figure, the conduction band has free electrons which have energies greater than E_c, the lowest energy level of the conduction band. An equal number of positively charged holes are present in the valence band of an intrinsic semiconductor material, which have energies smaller than E_v, the highest energy level of the valence band. The energy difference between the conduction band and the valence band is called the forbidden band which is devoid of any charge. The spread of this band is called the band-gap of the semiconductor material and is given by:

$$\text{Band gap,} = E_g = E_c - E_v \tag{3.14}$$

The recombination of an electron and a hole take place with the release of energy which is equal to the initial difference in the energies of the two charged particles. The released energy is given out in the form of radiation. If the frequency of the radiation falls within the visible range, a photon is said to be emitted and the process is known as Radiative Recombination. The energy of the emitted photon is equal to the product of the frequency of the radiation and the Planck's constant (6.626×10^{-34} Js). If the recombination does not result in the emission of a photon, the process is known as Non-Radiative Recombination. One should note that light, here, is treated in accordance to the quantum model which states light as a collection of photons. In our earlier discussion we had been using the ray and the wave model of light. So, in a semiconductor material, a part of the electron-hole recombination are radiative and the other part is not. The ratio of the number of radiative recombination to the total number of recombination in a semiconductor material is its radiation efficiency.

A semiconductor material is classified into two types on the basis of its energy band diagram in the energy-momentum space. These two categories of semiconductor materials are known as the direct and the indirect band-gap semiconductor materials. The figure below shows a schematic representation of the energy-band diagrams of these two types of semiconductors.

3.8.1. Direct band gap semiconductor

When the top of the valence band and the bottom of the conduction band have same value of momentum it is called as direct band gap semiconductor as shown in Fig. 3.20 (a).

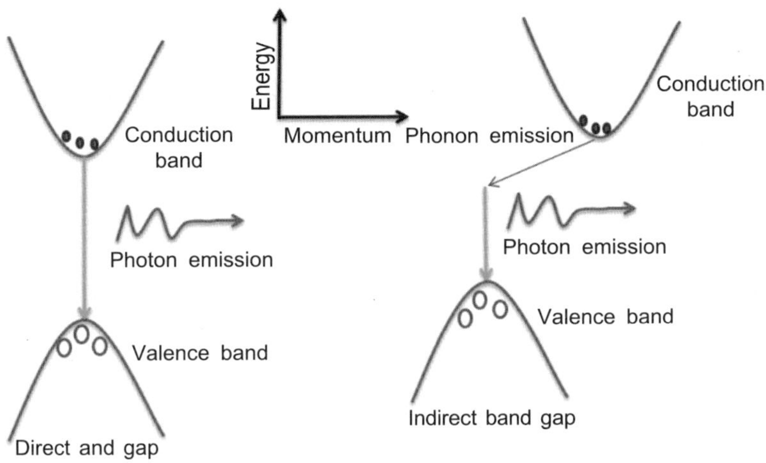

Fig. 3.20. (a) Direct and (b) indirect band gap.

A direct band-gap (DBG) semiconductor is one in which the maximum energy level of the valence band aligns with the minimum energy level of the conduction band with respect to momentum as shown in Fig. 3.20(a). On the other hand, if the two levels are misaligned with respect to momentum, the semiconductor is called as indirect band-gap (IBG) semiconductor (Fig. 3.20 (b)). In a DBG semiconductor, a direct recombination takes place with the release of the energy equal to the energy difference between the recombining particles. But in case of a IBG semiconductor, due to a relative difference in the momentum, first, the momentum is conserved by release of energy and only after the both the momenta align themselves, a recombination occurs accompanied with the release of energy. The probability of a radiative recombination, hence, in case of IBG semiconductor is much less in comparison to that in case of DBG semiconductors. Hence, the efficiency factor of a DBG semiconductor is much more than that of a IBG semiconductor. That is the reason why DBG semiconductors are always preferred over IBG for making optical sources. The two well-known intrinsic semiconductors, Silicon and Germanium are both IBG semiconductors and, hence cannot be used to manufacture optical sources. The most thoroughly investigated and studied DBG semiconductor material is Gallium Arsenide (GaAs). It has a high efficiency factor which means that a large portion of the recombinations in GaAs are radiative in nature. Apart from GaAs, there are also other materials which have direct band-gap nature. Most of them are the alloys of type III and type IV elements of the periodic table.

Once a suitable material for the optical source has been selected, the next work is to study the characteristics of the radiations emitted by it. This includes the ascertaining of the wavelength of the emitted radiation too. To understand that, let us have a look into an analogical situation of radiative recombination between two energy levels E_1 and E_2. The figure below shows the situation of a radiative recombination taking place between two energy levels, which in case of a semiconductor, may represent the maximum energy level of the valence band (i.e. E_1=E_v) and the minimum energy level of the conduction band (i.e. E_1=E_c). When a recombination of an electron in E_2 takes place with a hole in E_1, the energy difference (E_2 - E_1) is released in the form of a radiation of energy. If the wavelength, λ of this radiation falls within the visible range (4,000Å to 8,000 Å), a photon is said to have been emitted. The energy of this photon is given by:

$$E = E_2 - E_1 = hv \tag{3.15}$$

Here 'h' is the Planck's constant and 'v' is the frequency of the photon. The frequency of the photon when expressed in terms of wavelength and the velocity of the photon, changes the equation (3.15) to the one shown below

$$E = E_2 - E_1 = \frac{hc}{\lambda} \tag{3.16}$$

The quantity in the numerator is a constant and so, we can conclude that the wavelength of the radiation emitted is inversely proportional to the energy difference between the two levels between which the recombination takes place. The quantity in the numerator of equation being a constant, the wavelength of a photon emitted by the recombination can be represented as:

$$\lambda(\mu m) = \frac{1.24}{E(in\,eV)} = \frac{1.24}{E_2 - E_1\,(in\,eV)} \tag{3.17}$$

In case of a semiconductor material, the energy difference in the denominator of equation (3.17) would correspond to an energy difference equal to the band-gap of the semiconductor material. For a material like GaAs, the band-gap energy is about 1.4 eV. The wavelength of the photon emitted would, hence, be 0.8 μm (800 nm). If we recall, this was the wavelength of the sources required for the first generation of optical communication.

Thus GaAs was the material used for manufacture of optical sources for first generation optical communications which took place at a relatively low-loss window at around 800 nm.

As the optical window shifted from the first generation to the second and the third generations, so did the technology of manufacture of optical sources. Different ternary and quaternary types of substances were prepared which could be excited to emit light of desired wavelengths, just by changing the composition of their constituents in proper proportions.

By varying the value of the mole-fraction ‘x’ in the above substances, they can be made to be either DBG or IBG. The bounds on the value of ‘x’ indicate the region in which these materials remain DBG in nature. So by choosing appropriate value of ‘x’, the band-gap of the materials can be made to differ in accordance to the empirical equations given within parentheses against each material. Due to this variable band-gap nature of these materials, they can be made to emit light of arbitrary wavelengths (by equation 3.15) ranging from about 920 nm to 1,650 nm. In fact, this range almost matches with the modern low-loss window ranging from about 1,300 nm to 1,600 nm. So, in practical applications today, the optical sources used are mostly of quaternary nature which can be made to emit light at a particular desired wavelength suitable as per the application.

The distribution of electrons (or holes) in the conduction (or valence) band depends upon two main factors:

1. Availability of energy levels in the energy band. In other words,

the density of energy levels in the energy band.

2. Probability that a particular energy level has been occupied by an electron (or hole).

If the energy density function for the conduction band is considered to be represented as $S_c(E_2)$, the density of energy levels in the conduction band is given by the following relation (E_2 is a general energy level in the conduction band):

$$S_c(E_2) = \frac{4\pi(2m_e)^{\frac{3}{2}}}{h^2}\sqrt{E_2 - E_c} \tag{3.18}$$

Here 'm_e' is the mass of an electron, 'h' is the Planck's constant and 'E_c' is the minimum energy level in the conduction band or in other words, the start of the conduction band energy levels. In a similar manner, the energy level density in the valence band can be represented by $S_v(E_1)$ and can be written as (E_1 is a general energy level in the valence band):

$$S_v(E_1) = \frac{4\pi(2m_h)^{\frac{3}{2}}}{h^2}\sqrt{E_v - E_1} \tag{3.19}$$

The quantity 'm_h' represents the effective mass of a hole and 'E_v' is the maximum energy level in the valence band or the end of the valence band energy levels. If the band-gap of the above material is 'E_g', then then $E_g = E_c - E_v$.

If we observe equation 3.18 and 3.19 carefully, we find that

1. When $E_2 = E_c$, $S_c(E_2) = 0$; this suggests the absence of any energy level at the edge of the conduction band. Similarly, when $E_v = E_1$, $S_v(E_1) = 0$; suggests the absence of any energy level at the edge of the valence band.
2. The density of energy levels increases as we move deeper into the energy bands. This increase is in order of the term under the radical sign.

The probability that a particular energy level is occupied by an electron is given by the Fermi-distribution. So, if the Fermi-energy level (E_F) in the semiconductor material is known, the probability mentioned above can be written as:

$$F(E) = \frac{1}{1 + e^{(E - E_F)}/kT} \tag{3.20}$$

Here, 'E'=energy level under consideration, 'k' =Boltzmann's constant, 'T'=Absolute temperature of the semiconductor material. For an intrinsic semiconductor material, the Fermi energy level is midway between the valence and the conduction bands. The probability of an energy level to be occupied by hole is actually the probability of that energy level being not occupied by an electron and so is given by $1 - F_{(E)}$.

The product of the Fermi-distribution and the energy level density in a particular energy band, thus, gives the electron (or hole) distribution in the energy band. Therefore, the distribution of electrons and holes in the conduction and the valence bands is given by:

$$1-F(E)=1-\frac{1}{1+e^{(E-E_F)}/kT} \tag{3.21}$$

For a radiative recombination to occur, we must have electrons and holes in the semiconductor material that are available for a recombination. The simplest way to generate electrons and holes is by forward biasing a p-n junction (which is usually called a diode). This situation is depicted in the fig. 3.19 below:

To construct a p-n junction, we dope an intrinsic semiconductor material with appropriate type of impurities and in appropriate proportions to form two distinct regions which contain excess electrons and excess holes. The part of the semiconductor which is excess in electrons, is called the n-type side and the part with excess holes is called the p-type side. Once this semiconductor material is formed, the junction between the p-type and the n-type regions is called as the p-n junction. Under the effect of the forward bias voltage, the excess carriers are repelled towards the depletion region where they recombine and emit the recombination energy in the form of radiation. To compensate for the recombined electron hole pairs, more number of electrons and holes flow in through the terminals of the battery into the two regions and thus current flows in the circuit.

When an intrinsic semiconductor material is doped with an impurity, the Fermi-level (E_F) of the semiconductor specimen shifts and this shift depends on the type of the impurity whether it's a donor or an acceptor impurity. Accordingly, the doped semiconductor would then be called as the n-type or the p-type semiconductor respectively. Hence, for a p-type material, the probability of a hole in the valence band or the probability of absence of an electron in the valence band is given by:

But for a p-type semiconductor material, the second term in the denominator is much less than unity. Therefore, expanding the denominator binomially and retaining only upto the first order terms, we have:

$$1-F(E) \approx e^{(E_1 - E_{Fp})}/kT \qquad (3.22)$$

In a similar manner, if we calculate the probability of an electron in the conduction band of an n-type semiconductor material, we find that the exponential term in the denominator of equation 3.22 is much greater than unity. So, the probability of an electron in the conduction band of an n-type semiconductor material is given by:

$$F(E) \approx e^{-(E_1 - E_{Fn})}/kT \qquad (3.23)$$

Equation 3.22 and 3.23 give the probability of a hole and an electron in the conduction and valence band, respectively, in a semiconductor material. The product of these two probabilities is, in fact, proportional to the probability of a radiative recombination in a semiconductor material (if we consider all the recombinations in the semiconductor material to be radiative in nature).

3.8.2. Indirect band gap semiconductor

When the maxima energy of the valence band lies at a

different value of momentum to the minima in the conduction band energy it is called as indirect band gap semiconductor:

In optical devices this difference is vital. It is a matter of fact that, a photon can provide the energy to produce an electron-hole pair.

Each photon of energy E has momentum $p = E\,c$, where c is the velocity of light. An optical photon has energy of the order of 10^{-19} J, and, since, $c = 3 \times 10^{8}$ ms^{-1}, a typical photon has a very small amount of momentum.

In a direct band gap semiconductor it is quite easy to produce an electron hole-pair for a photon of energy E_g (where E_g is the band gap energy) since the electron does not need to be given very much momentum. However, in an indirect semiconductor an electron must also endure a noteworthy change in its momentum for a photon of energy E_g to produce an electron-hole pair. It occurs when an electron interacts with both the photon to gain energy and a lattice vibration called a phonon to either gain or lose momentum.

The indirect process is slow because it requires three things to meet in order to proceed those are an electron, a photon and a phonon. This is analogous to chemical reactions, because it is obvious in a particular reaction step that, a reaction between two molecules will proceed at a much greater rate than a reaction involving three molecules.

Similar principle is applicable to recombination of electrons and holes to produce photons. For a direct band gap semiconductor the recombination process is much more efficient than for an indirect band gap semiconductor, in which the process must be mediated by a phonon.

This is the main reason why gallium arsenide and related direct band gap semiconductors are being used to make optical devices such as light emitting diodes (LEDs) and semiconductor lasers and silicon is not used due to its indirect band gap.

3.9. Generation and recombination

A process by which both carriers electrons and holes annihilate each other and further electrons occupy through single or multiple steps while the empty state is occupied by a hole is called as recombination of electrons and holes. Both carriers ultimately vanish in the process. In this process the energy difference between the initial and final state of the electron is released. This leads to one possible classification of the recombination processes. When this energy is emitted in the form of a photon such recombination is called as radiative recombination. In the case of non-radiative recombination, it is passed on to one or more phonons and in the case of Auger recombination it is given off in the form of kinetic energy to another electron. Another classification scheme considers the individual energy levels and particles involved. These different processes are further illustrated with Fig. 3.21.

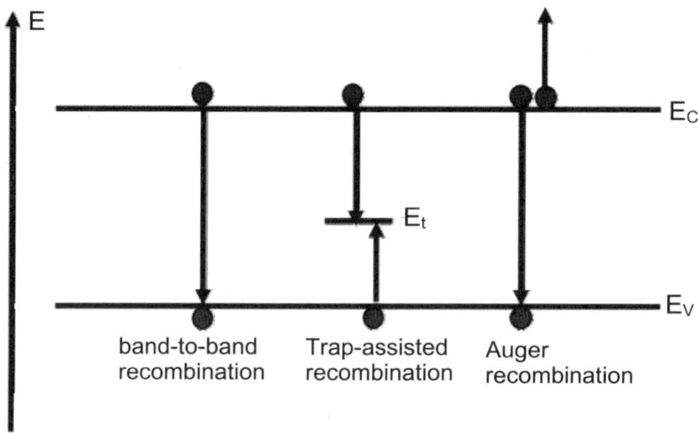

Fig. 3.21. Carrier recombination mechanisms in semiconductors.

Movement of an electron from its conduction band state into the empty valence state associated with the hole is band-to-band recombination. In direct band gap semiconductors this band-to-band transition is typically also a radiative transition.

When an energy level within the band gap causes the presence of foreign atom or structural defect it is called as trap. Trap-assisted recombination occurs when an electron falls into such "trap". The electron occupying the trap, in a second step, moves into an empty valence band state, thereby completing the recombination process. One can foresee this process as a two-step transition of an electron from the conduction band to the valence band or as the annihilation of the electron and hole, which meet each other in the trap. This process is referred as Shockley-Read-Hall (SRH) recombination.

When an electron and a hole recombine in a band-to-band

transition such recombination is called as Auger recombination, but in this case the resulting energy is given off to another electron or hole. Auger recombination is differently from band-to-band recombination because of the involvement of a third particle affects the recombination rate.

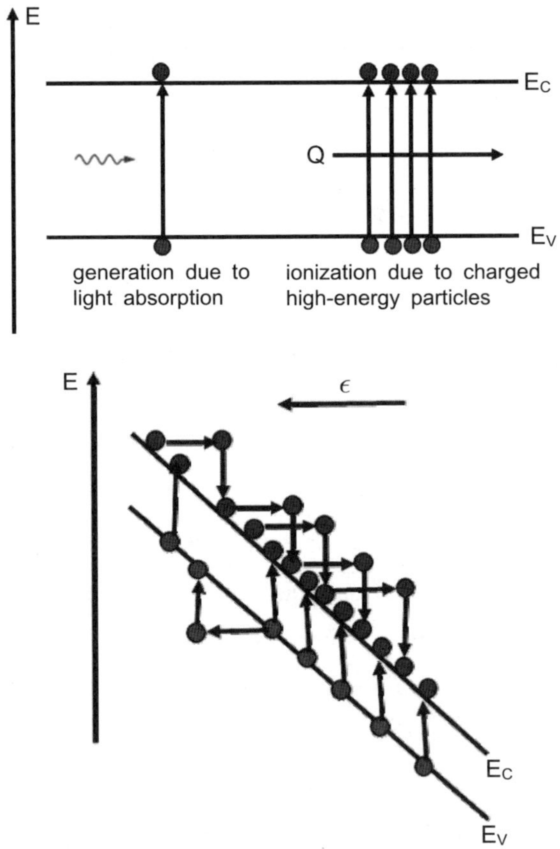

Fig. 3.22. (a) Carrier generation due to light absorption and ionization due to high-energy particle beams (b) Impact ionization and avalanche multiplication of electrons and holes in the presence of a large electric field.

Carrier generation occurs reversing each of these recombination mechanisms. A single expression will be used to describe recombination as well as generation for each of the above mechanisms.

In addition, there are generation mechanisms, which do not have an associated recombination mechanism, such as generation of carriers by light absorption or by a high-energy electron/particle beam. These processes are referred to as ionization processes. Impact ionization, which is the generation mechanism associated with Auger recombination, also belongs to this category. The generation mechanisms are illustrated with Fig. 3.22.

Carrier generation due to light absorption occurs if the photon energy is large enough to raise an electron from the valence band into an empty conduction band state, thereby generating one electron-hole pair. The photon energy needs to be larger than the band gap energy to satisfy this condition. The photon is absorbed in this process and the excess energy, E_{ph} - E_g, is added to the electron and the hole in the form of kinetic energy.

Carrier generation or ionization due to a high-energy beam consisting of charged particles is similar except that the available energy can be much larger than the band gap energy so that multiple electron-hole pairs can be formed. The high-energy particle gradually loses its energy and eventually stops. This generation mechanism is used in semiconductor-based nuclear particle counters. As the number of ionized electron-hole pairs varies with the energy of the particle, one can also use such detector to measure the particle energy.

3.9.1. Simple recombination-generation model

The excess energy is given off to generate an electron-hole pair through a band-to-band transition. This generation process causes avalanche multiplication in semiconductor diodes under high reverse bias: As one carrier accelerates in the electric field it gains energy. The kinetic energy is given off to an electron in the valence band, thereby creating an electron-hole pair. The resulting two electrons can create two more electrons which generate four more causing an avalanche multiplication effect. Electrons as well as holes contribute to avalanche multiplication.

A simple model for the recombination-generation mechanisms states that the recombination-generation rate is proportional to the excess carrier density. It acknowledges the fact that no net recombination takes place if the carrier density equals the thermal equilibrium value. The resulting expression for the recombination of electrons in a p-type semiconductor is given by:

$$U_n = R_n - G_n = \frac{n_p - n_{p0}}{\tau_n} \tag{3.24}$$

and similarly for holes in an n-type semiconductor:

$$U_p = R_p - G_p = \frac{p_n - p_{n0}}{\tau_p} \tag{3.25}$$

where the parameter can be interpreted as the average time after which an excess minority carrier recombines.

We will show for each of the different recombination mechanisms that the recombination rate can be simplified to this form when applied to minority carriers in a “quasi-neutral” semiconductor. The above expressions are therefore only valid under these conditions. The recombination rates of the majority carriers equals that of the minority carriers since in steady state recombination involves an equal number of holes and electrons. Therefore, the recombination rate of the majority carriers depends on the excess-minority-carrier-density as the minority carriers limit the recombination rate.

Recombination in a depletion region and in situations where the hole and electron density are close to each other cannot be described with the simple model and the more elaborate expressions for the individual recombination mechanisms must be used.

3.9.2. Band-to-band recombination

Band-to-band recombination depends on the density of available electrons and holes. Both carrier types need to be available in the recombination process. Therefore, the rate is expected to be proportional to the product of n and p. Also, in thermal equilibrium, the recombination rate must equal the generation rate since there is no net recombination or generation. As the product of n and p equals n_i^2 in thermal equilibrium, the net recombination rate can be expressed as:

$$U_{b-b} = b(np - n_i^2) \qquad (3.26)$$

where b is the bimolecular recombination constant.

3.9.3. Trap assisted recombination (Shockley-Read-Hall)

The net recombination rate for trap-assisted recombination is given by:

$$U_{SHR} = \frac{pn - n_i^2}{p + n + 2n_i \cosh\left(\frac{E_i - E_t}{kT}\right)} N_t v_{th} \sigma \tag{3.27}$$

The derivation of this equation is beyond the scope of this text.

This expression can be further simplified for p >> n to:

$$U_n = R_n - G_n = \frac{n_p - n_{po}}{\tau_n} \tag{3.28}$$

And for n >> p to:

$$U_p = R_p - G_p = \frac{p_n - p_{no}}{\tau_p} \tag{3.29}$$

Were

$$\tau_n = \tau_p = \frac{1}{N_t v_{th} \sigma} \tag{3.30}$$

3.10. Surface recombination

$$U_{S,SHR} = \frac{pn - n_i^2}{p + n + 2n_i \cosh\left(\frac{E_i - E_{s_t}}{kT}\right)} N_{st} v_{th} \sigma_s \tag{3.31}$$

This expression is almost identical to that of Shockley-Hall-Read recombination. The only difference is that the recombination is due to a two-dimensional density of traps, N_{st}, as the traps only exist at the surface or interface.

This equation can be further simplified for minority carriers in a quasi-neutral region. For instance for electrons in a quasi-neutral p-type region $p >> n$ and $p >> n_i$ so that for $E_i = E_{st}$, it can be simplified to:

$$U_{S,n} = R_{s,n} - G_{s,n} = v_s(n_p - n_{p0}) \tag{3.32}$$

where the recombination velocity, vs, is given by:

$$v_s = N_{st} v_{th} \sigma_s \tag{3.33}$$

3.10.1. Auger recombination

Auger recombination involves three particles: an electron and a hole, which recombines in a band-to-band transition and gives off the resulting energy to another electron or hole. The expression for the net recombination rate is therefore similar to that of band-to-band recombination but includes the density of the electrons or holes, which receive the released energy from the electron-hole annihilation:

$$U_{Auger} = \Gamma_n n(np - n_i^2) + \Gamma_p p(np - n_i^2) \tag{3.34}$$

The two terms correspond to the two possible mechanisms.

References

1. P. Y. Yu and M. Cardona, Fundamentals of Semiconductors: Physics and Materials Properties (Graduate Texts in Physics), Springer-Verlag, Berlin, 2010.
2. B. G. Streetman and S. M. Banerjee, Solid State Electronic Devices, PHI Learning Private Ltd, New Delhi, 2009.
3. C. Kittel, Introduction to Solid State Physics, John Wiley and Sons, Inc, Hoboken, NJ, 2005.
4. M. A. Wahab, Solid State Physics: Structure and Properties of Materials, Alpha Science International Ltd, Oxford, 2005.
5. R. Enderlein and N. J. M. Horing, Fundamentals of Semiconductor Physics and Devices, World Scientific Publishing Co. Pvt. Ltd., Singapore, 1999.
6. http://www.virginia.edu/, Class Notes
7. http://www.chemhume.co.uk/, A Level Chemistry
8. http://ecee.colorado.edu/, B. Van Zeghbroeck, 2011.

Chapter 4
P-N junction and analysis

4.1. Introduction

In 1939, while experimenting on cat whiskers for detection of radio waves the scientist Russel Ohl from Bell laboratories, found highly improved and uniform electrical properties for melt fabricated silicon material. He understood that this improvement was due to the impurities in silicon which created the p-n junction that he accidently obtained during his experiments. Later on the accidently invented p-n junction became a heart/basic building block of almost all electronic devices like solar cells, transistors, radios, televisions, power supplies etc. We have already discussed on the p-type and n-type semiconductors in details in earlier chapters. The p-type semiconductor is formed on adding trivalent impurity in the intrinsic semiconductor while n-type semiconductor is formed on addition of pentavalent impurity to intrinsic semiconductor.

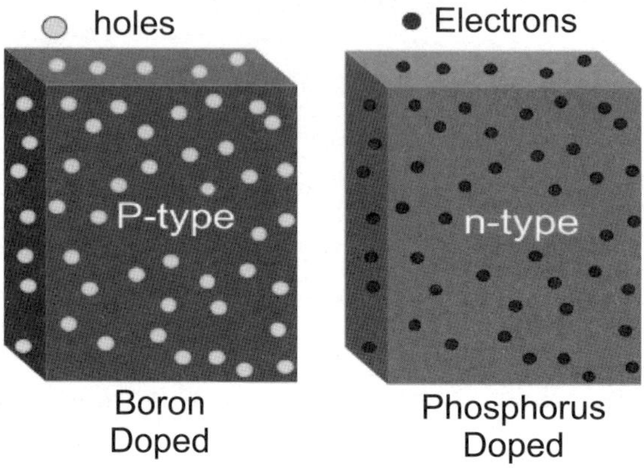

Fig. 4.1. p-type and n-type semiconductor with holes and electrons as majority carriers respectively.

The p-type semiconductor has majority of holes, while n-type semiconductor possesses majority of electrons as shown in Fig. 4.1. Even though p-type semiconductor has large number holes it is electrically neutral at the thermodynamic equilibrium. Also n-type semiconductor is electrically neutral at thermodynamic equilibrium even it possesses large number of electrons. It is because the holes in p-type semiconductor and electrons in n-type semiconductor are provided by trivalent and pentavalent atoms respectively which are electrically neutral. When such a p and n-type semiconductors comes together through a proper scheme, it is the interface that forms the part of interest, where p-n junction is formed as shown in Fig. 4.2 below.

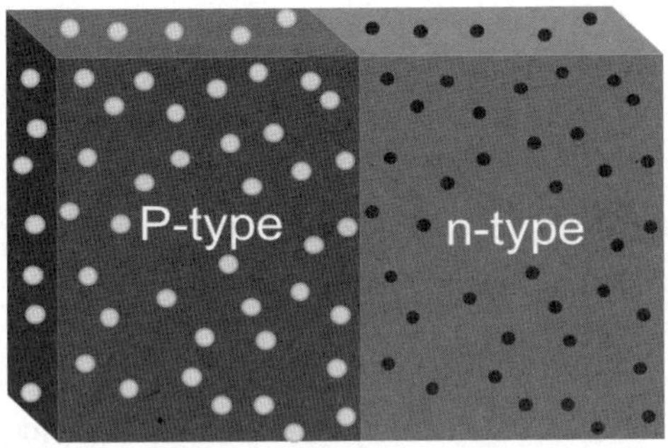

Fig. 4.2. p-type and n-type semiconductor in contact. (The diffusion and recombination process has not yet occurred.)

The properties of p-n junction are very useful and different as compared to individual p and n-type semiconductor which, makes it a very popular for its applications in various fields like diodes, transistors, integrated chips etc. Almost all semiconductor devices consists of single crystal semiconductor which offers wide range of essentially nonlinear conductivity effects by virtue of junctions. Where the junction is the difference in electric field and carrier density in semiconductor; created by different impurity doping density forming a different semiconducting regions. All semiconductor devices can be understood through the study of p-n junction. Where the p-n junction is the two semiconducting regions in physical contact with each other; one is having excess of holes and other having excess of electrons.

4.2. Formation of p-n junction/unbiased p-n junction

The two important phenomenon taking place during the

formation of p-n junction is diffusion and drift. We know that p-type semiconductor has high concentration of holes and n-type semiconductor has high concentration of electrons. Even at room temperature there is the thermal energy associated with materials which causes thermal perturbations or thermal agitations in material. During the formation of p-n junction, the combination of prevalent concentration gradient and thermal agitations causes holes from p-side to diffuse to n-side and electrons from n-side to diffuse to p-side as shown in Fig 4.3. This motion of charge carriers across the junction leads to diffusion current. To understand this process of diffusion, just consider a simple examples of spreading of an ink in a water or a fragrance in air which has a similar analogy. This process of spreading out due to random velocity and scattering is known as diffusion.

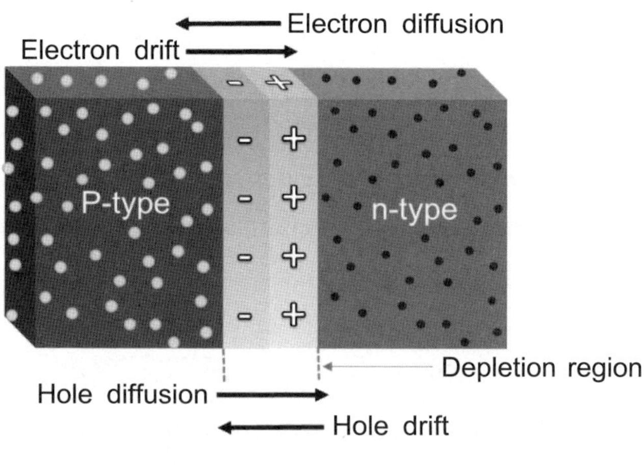

Fig. 4.3. Diffusion of holes from p-side to n-side and electrons from n-side to p-side due to thermal agitations.

When an electron from n-side diffuses to p-side to get to be annihilated, it leaves behind an immobile positively charged ionized donor atom (that has donated an electron) on n-side. An ionized donor atom is immobile as it is bonded to the surrounding atoms. As the electrons continues to diffuse from n-side to p-side, a layer of positively charged immobile ionized donor atom forms a positive space charge region on the n-side of junction. Similarly when a hole from p-side diffuses to n-side, it leaves behind an immobile negatively charged ionized acceptor atom (that has accepted an electron) on p-side. An ionized acceptor atom is also immobile as it is bonded to the surrounding atoms. As the holes continues to diffuse from p-side to n-side, a layer of negatively charged immobile ionized acceptor atom forms a negative space charge region on the p-side of junction. The space charge regions on either side of the junction together known as the depletion region, as it is depleted of free charge carriers. It is depleted because the free charge carriers in this regions have been annihilated. The width of depletion region is of the order of one-tenth of a micrometer. Due to the positive space charge and negative space charge on n-side and p-side of junction respectively, an electric field directing from positive charge to negative charge gets developed. This electric field causes motion of charge carriers, in which minority charge carrier holes from n-side moves to p-side, whereas the minority carrier electrons from p-side moves to n-side. This motion of charge carriers under influence of an electric field is known as drift. It leads to drift current which is opposite in direction to the diffusion current.

Initially the drift current is small and diffusion current is high. Further as the diffusion process continues, the width of space-charge region extends on both side of junction, it enhances the electric field strength as well as drift current. This process continues until the drift current equals the diffusion current. In this way the formation p-n junction takes place. In p-n junction at equilibrium there is no net current. The potential difference is developed across the junction of two regions (as shown in Fig. 4.4) due loss of electron from n-side and the gain of electron by the p-side of junction. The polarity of developed potential is such that the further diffusion of carriers is prevented so that the equilibrium condition exists.

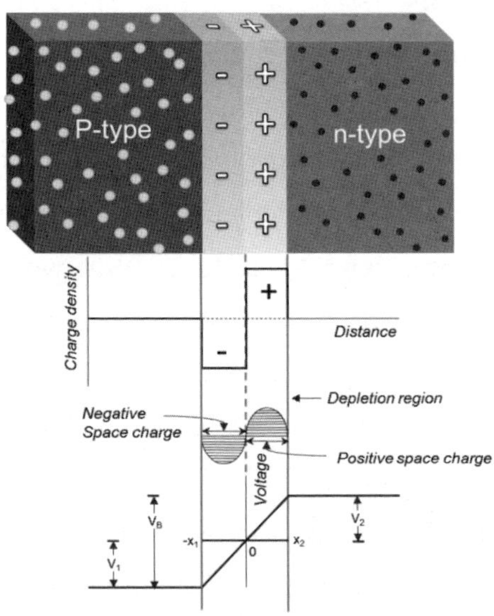

Fig. 4.4. p-n junction under equilibrium and formation of barrier potential.

Fig. 4.4 depicts the p-n junction at equilibrium and the electrostatic potential difference across the junction. This electrostatic potential is called as barrier potential as it discourages the diffusion of majority charge carriers across the junction. The potential barrier for silicon p-n junction and germanium p-n junction is 0.7 V and 0.3 V respectively. The voltage developed across the junction can be represented by virtual battery having voltage V_B. The potential difference occurs across the space charge region only outside this region on both side junctions the semiconductor is still neutral.

4.3. Semiconductor p-n junction diode

When a metallic contacts are provided at the end of p-n junction it forms p-n junction diode. Through these metallic contacts we can apply external voltage and thereby we can vary equilibrium barrier potential. p-n junction diode is two terminal device and it is symbolically represented as shown in Fig. 4.5. The direction of arrow indicates the direction of conventional current when diode is under forward bias.

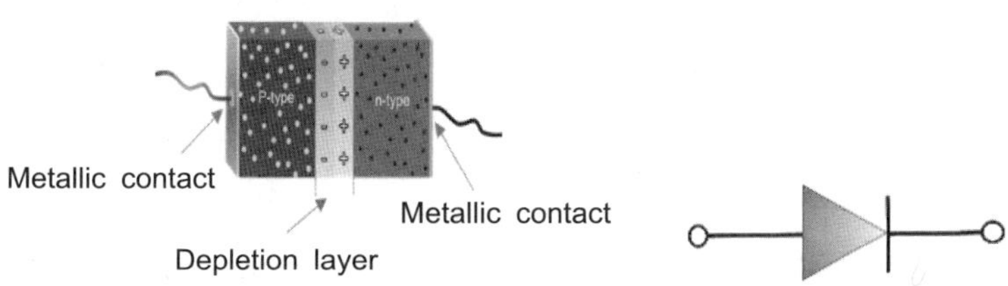

Fig. 4.5. p-n junction diode and its symbolic representation.

The Fig. 4.5 Shows the p-n junction diode at equilibrium condition. There will be no net movement of charge carriers unless and until external field is applied across the p-n junction. The electric field can be applied to p-n junction in two ways (1) forward biasing (II) reverse biasing.

4.3.1. P-N junction diode under forward bias

The p-n junction diode is said to be forward biased, when an external voltage V is applied across the diode with a polarity such that majority carriers from both sides are driven towards the junction. In simple way for forward biasing the positive terminal of battery is connected to p-side of semiconductor diode and the negative terminal of battery is connected to n-side of diode as shown in Fig 4.6. The resistance to the applied voltage is very high at the depletion region as compared to resistance at n-side and p-side. Hence the most of applied voltage drop occurs across the depletion region, whereas the voltage drop across the n-side and p-side is very negligible. The applied forward voltage establish an electric field which is opposite in direction to the built-in potential V_0. Hence on application of forward potential the width of depletion layer decreases and barrier height reduces. The effective barrier height under forward bias is (V_0 - V). If the applied voltage is small, the barrier potential will be reduced only slightly below the equilibrium value, and only a small number of carriers in the material - those that happen to be in the uppermost energy levels - will possess enough energy to cross the junction. So the current will be small. If we increase the applied voltage significantly, the barrier height will be reduced and more number of carriers will have the required energy. Thus the current increases. Due to the applied voltage, electrons from n-side cross the depletion region and reach p-side (where they are minority carriers). Similarly, holes from p-side cross the junction and reach the n-side (where they are minority carriers).

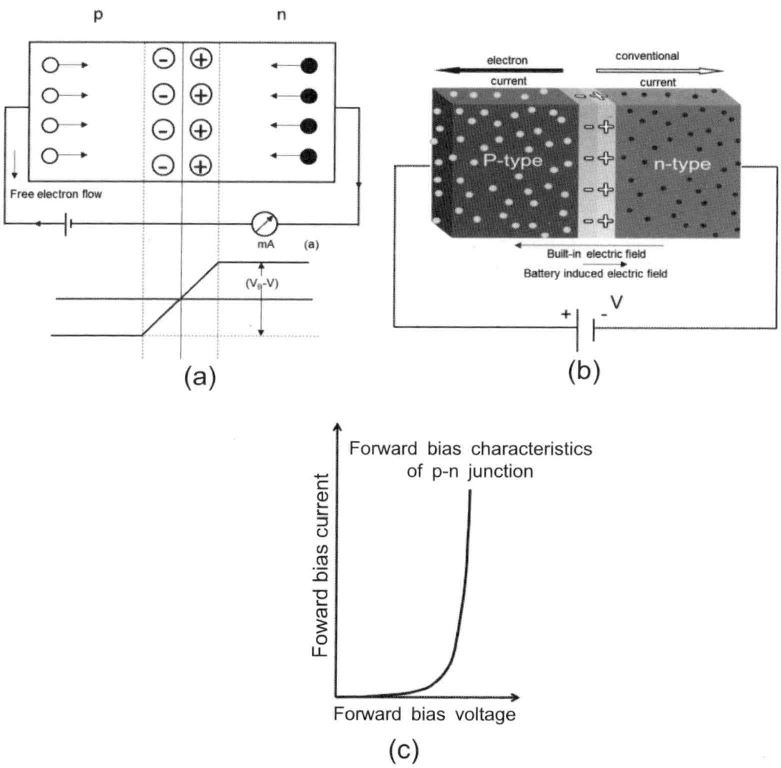

Fig. 4.6. (a) & (b) p-n junction under forward bias condition and (c) its I-V characteristics.

This process under forward bias is known as minority carrier injection. At the junction boundary, on each side, the minority carrier concentration increases significantly compared to the locations far from the junction. Due to this concentration gradient, the injected electrons on p-side diffuse from the junction edge of p-side to the other end of p-side. Likewise, the injected holes on n-side diffuse from the junction edge of n-side to the other end of n-side (Fig. 4.6). This motion of charged carriers on either side gives rise to current. The total diode forward current is sum of hole diffusion current and conventional current due to electron diffusion. The magnitude of this current is usually mA.

4.3.2. P-N junction diode under reverse bias

When an external voltage (V) is applied across the diode such that n-side is positive and p-side is negative, it is said to be reverse biased (Fig. 4.7). The applied voltage mostly drops across the depletion region. The direction of applied voltage is same as the direction of barrier potential. As a result, the barrier height increases and the depletion region widens due to the change in the electric field. The effective barrier height under reverse bias is (V_0 + V), (Fig. 4.7). This suppresses the flow of electrons from n-side to p-side and holes from p-side to n-side. Thus, diffusion current, decreases enormously compared to the diode under forward bias. The electric field direction of the junction is such that if electrons on p-side or holes on n-side in their random motion come close to the junction, they will be swept to its majority zone. This drift of carriers gives rise to current. The drift current is of the order of a few μA. This is quite low because it is due to the motion of carriers from their minority side to their majority side across the junction. The drift current is also there under forward bias but it is negligible (μA) when compared with current due to injected carriers which is usually in mA. The diode reverse current is not very much dependent on the applied voltage.

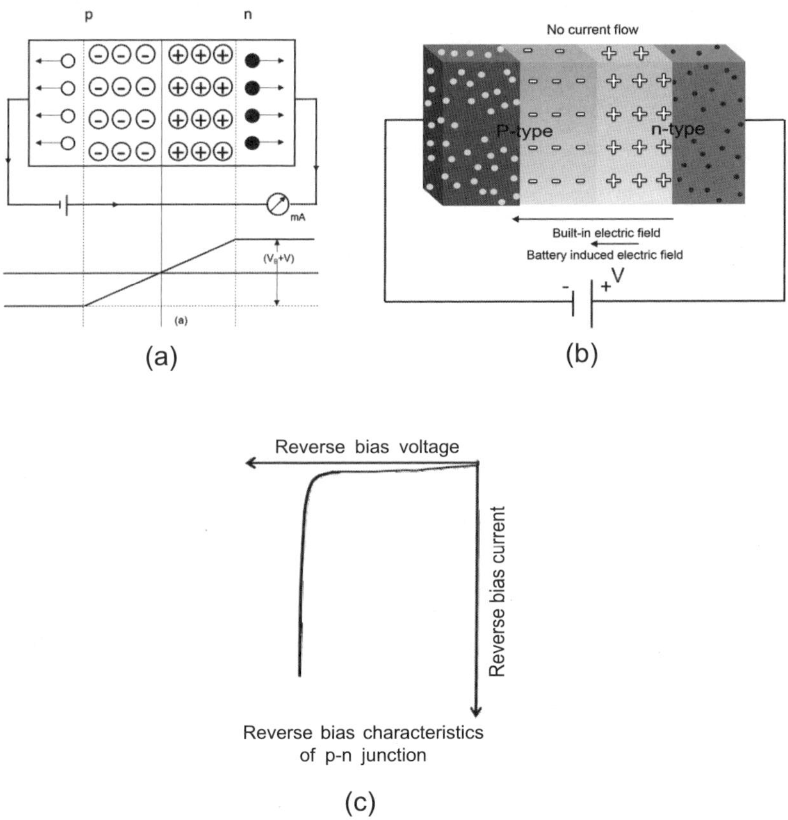

Fig. 4.7. (a) and (b) p-n junction under reverse bias condition and (c) its I-V characteristics.

Even a small voltage is sufficient to sweep the minority carriers from one side of the junction to the other side of the junction. The current is not limited by the magnitude of the applied voltage but is limited due to the concentration of the minority carrier on either side of the junction. The current under reverse bias is essentially voltage independent upto a critical reverse bias voltage, known as breakdown voltage (V_{br}). When V = V_{br}, the diode reverse current increases sharply. Even a slight increase in the bias voltage causes large change in the current. If the reverse current is not limited by an external circuit below the rated value (specified by the manufacturer) the p-n junction will get destroyed. Once it exceeds the rated value, the diode gets destroyed due to overheating. This can happen even for the diode under forward bias, if the forward current exceeds the rated value. The circuit arrangement for studying the I-V characteristics of a diode, (i.e., the variation of current as a function of applied voltage) are shown in Fig. 4.7 (a) and (b). The battery may be connected to the diode through a potentiometer (or reheostat) so that the applied voltage to the diode can be changed. For different values of voltages, the value of the current is noted. A graph between V and I is obtained as in Fig. 4.7 (c). Note that in forward bias measurement, we use a milli-ammeter since the expected current is large (as explained in the earlier section) while a micrometer is used in reverse bias to measure the current. You can see in Fig. 4.6 (c) that in forward bias, the current first increases very slowly, almost negligibly, till the voltage across the diode crosses a certain value. After the

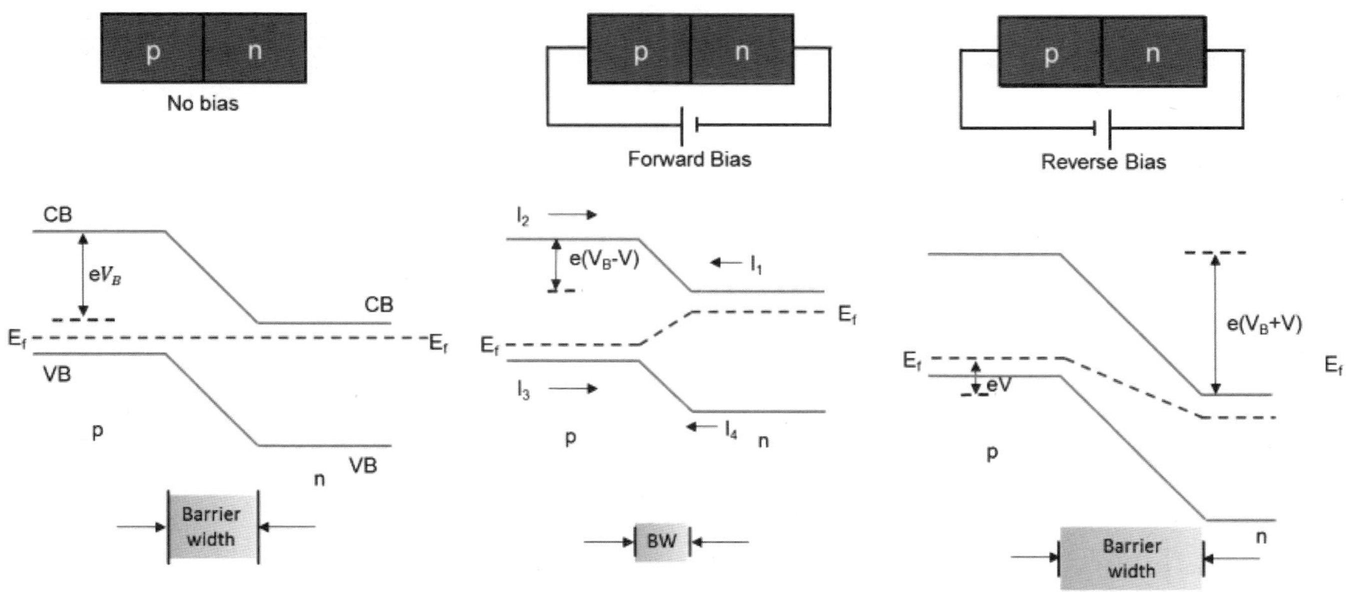

Fig. 4.8. Band diagram of unbiased, forward biased and reverse biased p-n junction.

characteristic voltage, the diode current increases significantly exponentially), even for a very small increase in the diode bias voltage. This voltage is called the threshold voltage or cut-in voltage (~0.2V for germanium diode and ~0.7 V for silicon diode).

4.4. Diode current equation

For a given p-n junction the total current density across it can be understood by taking in to consideration four constituent components. Let us consider that the first component is the current density due to minority carriers I_1. The minority electrons in the conduction band of p-region flow to conduction band of n-region. As per the convention the current flow in an opposite direction to the flow of negatively charged electrons. In present case the electrons are moving from the p-region to the n-region and hence the current I_1 flow from the n-region to the p-region i.e. I_1 flow from right to left. The second constituent component I_2 is the current density due to majority carriers. The majority charge carriers in the conduction band of the n-region i.e. electrons, can move in to the conduction band of the p-region giving rise to current I_2 flowing from left to right.

Similarly the majority charge carriers in the valence band in the p-region is holes. These holes moves from the valence band of p-region to valance band of n-region and is responsible to the current density I_3 having direction from left to right. At the same time the minority charge carriers in the valence band of n-region which is holes can flow to the valence band of the p-region responsible to a current density I_4 having direction from right to left.

Therefore the total current density I from left to right is given by

$$I = (I_2 + I_3) - (I_1 + I_4) \tag{4.1}$$

Now that we have defined the individual components of current density we try to evaluate each of them. The current density I_1 represents the flow of electrons down the energy hill from the p-side to the n-side. As discussed earlier this current is due to the minority electrons in the p-side. Being minority charge carriers their concentration is small and hence they contribute to a very small current. Now knowing that the potential is applied only across the barrier region and not in the bulk of p-n junction, we can imagine that the motion of these minority electrons in p-side is totally governed only by diffusion. It is only a matter of diffusion up to the junction boundary, after which electron will be easily carried through the junction down the potential hill into the n-side. This process is independent of applied voltage and the electrons always sees an easy path to traverse down the potential hill while moving from left to right, no matter whether its reverse biased or forward biased. Thus I_1 is always remain small but practically constant and called as saturation current density. Now the current I_2 is in existence due to the motion of electrons (which is the majority charge carrier from the conduction band) in n-region to cross the barrier with a potential $V_T = (V_r - V)$. The electrons having the energy sufficient enough to surpass this potential barrier V_T will cross the junction to the left. So the current density I_2 can be given as:

$$I_2 + C_1 exp\left(\frac{-eV_T}{k_B T}\right) \tag{4.2}$$

Where C_1 is a constant. Now analogous to the electron in the conduction band of n-region, one can give exact similar argument for holes in the valance band of p-region. Hence I_4 forms a saturation current density which is independent of applied potential and I_3 is the current due to flow of hole climbing up the potential hill. Hence the following equation:

$$I_3 = C_2 exp\left(\frac{-eV_T}{k_B T}\right) \tag{4.3}$$

Where C_2 is another constant.

Consider there is no voltage applied across the junction (V=0), now in this situation the overall current density for both holes and electrons is separately zero. As there is no accumulation of charges in either band under equilibrium, the current flow in one band is independent of other. And thus by applying the equilibrium condition i.e. $V = 0$, C_1 and C_2 can be evaluated. Hence, if $V = 0$, $V_T = V_B$, $I_2 = I_1$ and $I_3 = I_4$ Substitute the value of I_2 and I_3 in equation (4.2) and (4.3), we get

$$I_1 = C_1 exp\left(\frac{-eV_B}{K_B T}\right) \text{ and } I_4 = C_2 \exp\left(\frac{-eV_T}{K_B T}\right)$$

Therefore

$$C_1 = I_1 exp\left[\frac{eV_B}{K_BT}\right]$$

$$C_2 = I_4 exp\left[\frac{eV_B}{K_BT}\right]$$

Substitute the values of C_1 and C_2 in equation (4.2) and (4.3), we get

$$I_1 = I_1 exp\left[\frac{e(V_B - V_T)}{K_BT}\right]$$

And
$$I_3 = I_4 exp\left[\frac{e(V_B - V_T)}{K_BT}\right]$$

Thus

$$I_2 - I_3 = \left\{(I_1 + I_4) exp\left[e\left(\frac{V_B - V_T}{K_BT}\right)\right]\right\}$$

Hence the total current density I is given by

$$I = (I_1 + I_4)\left\{exp\left[e\left(\frac{V_B - V_T}{K_BT}\right)\right] - 1\right\}$$

Here, the total saturation current density $I_0 = (I_1 + I_4)$ and the applied voltage $V = (V_B - V_T)$.

Thus the current density equation becomes,

$$I = I_0\left[exp\left(\frac{eV}{K_BT}\right) - 1\right] \tag{4.4}$$

This equation is known as the diode current equation. If the applied voltage biases the junction in forward direction V is positive and if the voltage biases the junction in reverse direction V is negative. The graphical representation of the diode equation (4.4) is as shown in Fig. 4.9. In the forward bias condition $V > K_BT/e$, and $\exp(eV/K_BT) >> 1$. Hence $I = I_0 \exp(eV/K_BT)$ which implies that current I increases exponentially with applied voltage. On the contrary, under a reverse bias condition, voltage $-V >> K_BT/e$ and $\exp(-eV/K_BT) << 1$. Hence practically $I = I_0$. Thus the current under the reverse bias practically remains constant at a minimal value of I_0, until the junction breaks down. This theoretical behavior is experimentally verified for diodes.

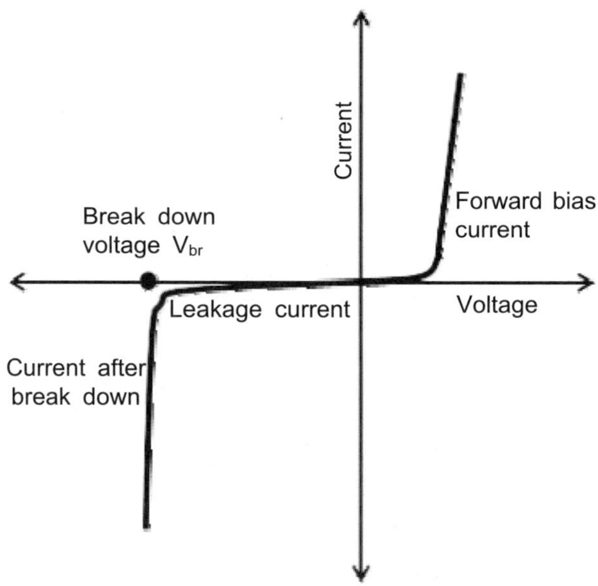

Fig. 4.9. I-V characteristics of p-n junction diode.

4.5. P-N junction under illumination / Photovoltaic effect in p-n junction diode

The photovoltaic effect refers to excitation of electrons in valence band of semiconductor to higher energy level due to absorption of incident photon's energy, and these photo excited electrons and their positive counterparts (holes) are able to generate an electrical current in an external circuit. The power cannot be generated with use of individual p or n type semiconductor as individual semiconductor cannot prevent the annihilation of generated electron hole pair. One needs an asymmetry in the electronic system to separate the generated positive and negative charges. This vital role of separating generated charges can be done with p-n junction. Also the device concepts like Schottky diodes or sandwich of a lowly doped semiconductor in between two different metal electrodes, and photo electrochemical device (i.e. creation of junction between solid semiconductor and liquid electrolyte containing redox couple) are some of the other possibilities. Whatever the structure they all havc, the one common thing all should have is, there should be an electric field present due to the implemented asymmetry in the system. This electric field can then act as a driving force to separate the photo-excited charges and have them collected by some outer electrodes. In this way, an electric current is generated in the external circuit. The working principle of photovoltaic cell can be easily understood with Fig. 4.10.

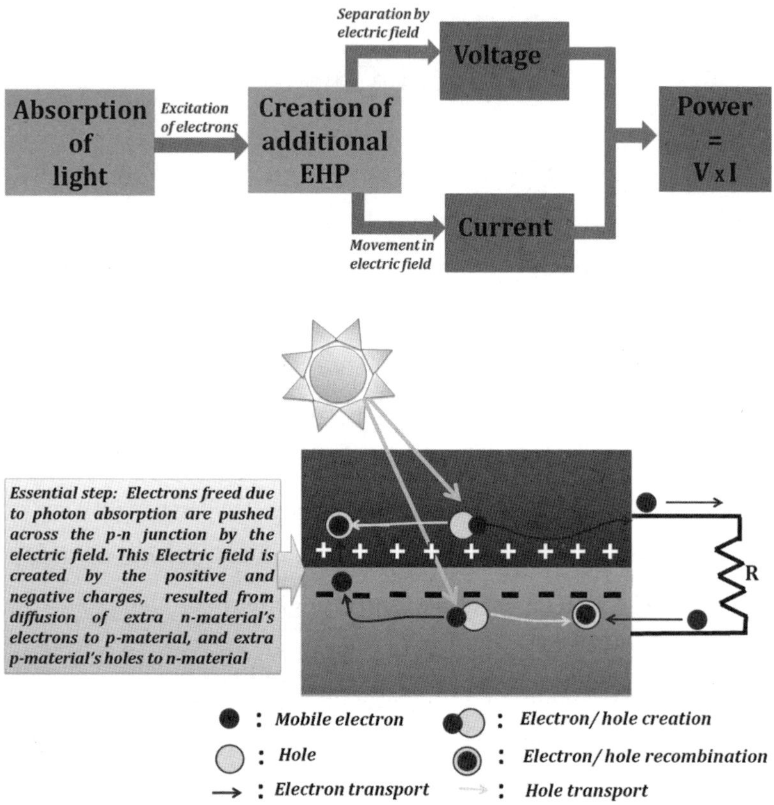

Fig. 4.10. Working principle of photovoltaic solar cell.

4.6. Types of junctions and dolar cells

There are various types of junctions made to exploit the remarkable electron transport features of semiconductors. One common and significant fact that remains more or less true for all these kind of junctions is the band offset observed by respective carriers, which emerges due to the difference in either the doping levels in two semiconductor forming the p-n junction or the band gaps of semiconductors forming the p-n junction. The carriers will face these band offsets or energy barriers which they have to cross by getting in to higher energy states, when transporting across the junctions.

4.6.1. Semiconductor-Semiconductor p-n junction (Homo-junction & Hetero-junction)

Basically a junction stands for an interface or a boundary where these two materials meet. In principle when a junction is formed in between two parts of same semiconductor materials having dissimilar electrochemical potential the junction is termed as a homo-junction. Specifically, a homo-junction is a junction in which p-type and n-type of a material having same band gap are in contact. Thus the p-n junction which we have discussed in sections above is a homo-junction having band gap structure as shown in Fig 4.11 (a) below. Using the same analogy, one can understand the nature of hetero-junction. A p-n junction is termed as a hetero-junction when the p-type and n-type of two different materials having dissimilar band gaps are brought in contact with each other. A typical example of energy band diagram of hetero-junction under a thermal equilibrium is as shown in Fig 4.11 (b).

To understand the band structure of semiconductor heterojunction we take the help of the term vacuum level and electron affinity.

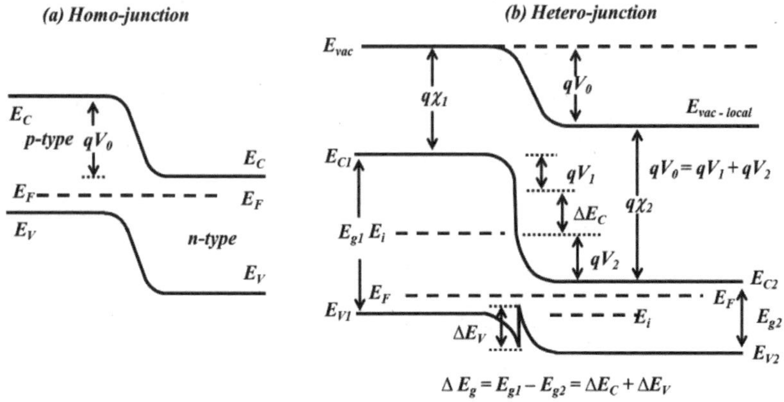

Fig. 4.11. Energy band diagram for (a) Homo-junction and (b) Hetero-junction semiconductor.

The vacuum level E_{vac} is defined as an energy level at which electrons is free from influence of any force, i.e. an electron is removed outside the semiconductor and is at the state of rest with respect to the semiconductor. The second term electron affinity ($q\chi$, where q is a unit charge) in any semiconductor is the energy required by an electron to move from conduction band minima to vacuum level. It has to be noted that the electron affinity is a characteristic of material and remains constant for that material.

In a heterojunction, when semiconductors that have different band gaps E_{g1} and E_{g2} are brought together, the band alignment is observed at the junction as the Fermi energy levels up at equilibrium. This band alignment is dictated by the Anderson electron affinity rule where the electron affinities ($q\chi_1$ and $q\chi_2$) influence the conduction band offset. The Anderson electron affinity rule states that the discontinuity at the conduction band edges is due to the difference of electron affinities of the constituent semiconductors. Hence the conduction band offset (ΔE_C) at heterojunction is equal to the difference in the electron affinity of the two semiconductors i.e. $\Delta E_C = q(\chi_2 - \chi_1)$ and the sum of discontinuity in the conduction band ΔE_C and valance band energy ΔE_V is equal to the difference in the band gap of individual semiconductors i.e. $(\Delta E_C + \Delta E_V) = \Delta E_g = (\Delta E_{g1} - \Delta E_{g2})$. Hence valance band offset $\Delta E_V = (\Delta E_{g1} - \Delta E_{g2}) - \Delta E_C$. Now since the electron affinity is a characteristic of material and does not change with the formation of heterojunction, the vacuum level also get realigned to a local vacuum level $E_{vac\text{-}local}$. The difference between the local and universal vacuum level is equal to the built in potential qV_0 as shown in Figure 4.11(b).

4.6.2. Metal-semiconductor junction solar cell (Schottky barrier)

In this kind of solar cells a transparent layer of metal film is usually evaporated on semiconductor surface. The interface between metal and semiconductor is known as Schottky junction plays active role in photogeneration. The Schottky junction creates built in potential across the bulk of semiconductor and the surface known as the Schottky barrier. An abrupt potential difference created due to difference in energy levels between conduction band of semiconductor and Fermi level of metal is known as Schottky barrier height. The Schottky junction solar cells can be prepared with number of materials; some of the popular examples are CdSe,

GaAs, NiO solar cells. The energy band diagram of Schottky barrier cell is shown below.

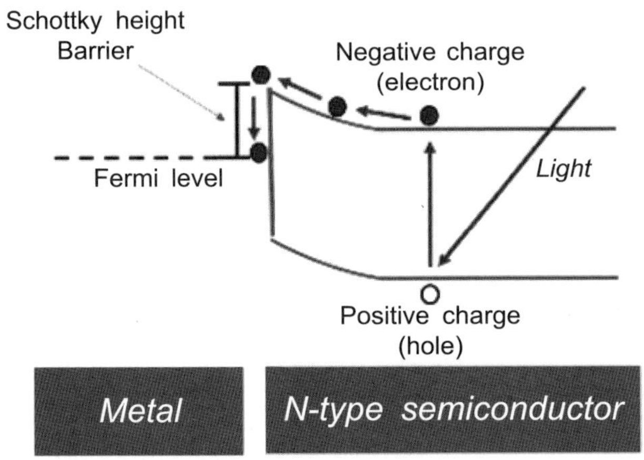

Fig. 4.12. Band diagram of Metal-semiconductor junction.

4.6.3. Metal-Insulator-Semiconductor (MIS) and Semiconductor-Insulator-Semiconductor (SIS) solar cell

A conventional homojunction or semiconductor heterojunction devices are fabricated from pure Si or other single crystal material which is an expensive affair. The cost of devices consisting of such junctions can be brought down only by a technological breakthrough in making these kind of p-n junctions and utilizing other polycrystalline materials or amorphous semiconductors as an alternative to single crystalline silicon. But these polycrystalline or amorphous materials are incompatible with conventional homojunction devices due to their high reverse saturation current.

Hence as an alternative, other options such as heterojuntions or surface barrier devices such as Schottky barrier were explored.

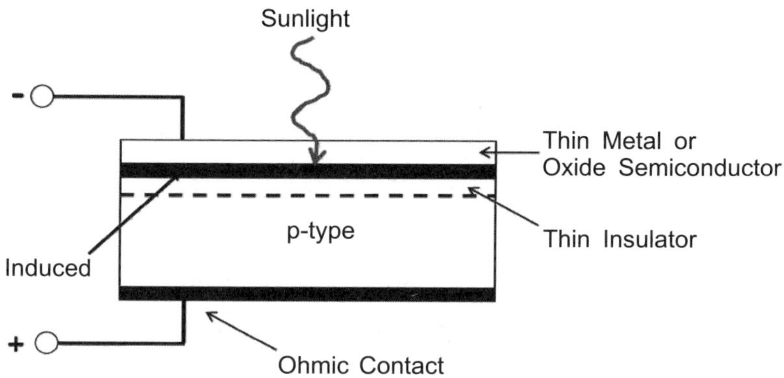

Fig. 4.13. Schematic diagram of MIS or SIS Junction.

It was found that a thin layer of the order of few tens of Angstroms can considerably increase the photovoltaic device performance. This lead the way towards formation of Metal-insulator-semiconductor or semiconductor-insulator-semiconductor junction devises. This is illustrated as in Fig. 4.13. Such a device depends on the thin interfacial layer between the top inducing contact and the base semiconductor. This interfacial layer is generally an oxide or a normal insulator in the bulk form.

4.7. Electrical characteristics of solar cells/ output parameters

4.7.1. Equivalent circuit of solar cell

Before studying the output parameters of solar cells, understanding the equivalent circuit of solar cell is quite important; as the relation between various out-put parameters of solar cells are established with reference of equivalent circuit.

The current I_L can be flown through load resistance R_L and voltage V_L can be developed across it by connecting the load to the terminals of solar cell. The magnitude of V_L and I_L rather than depending on nature of load resistance it depends on the properties of diode and the photo-generated current I_{ph}.

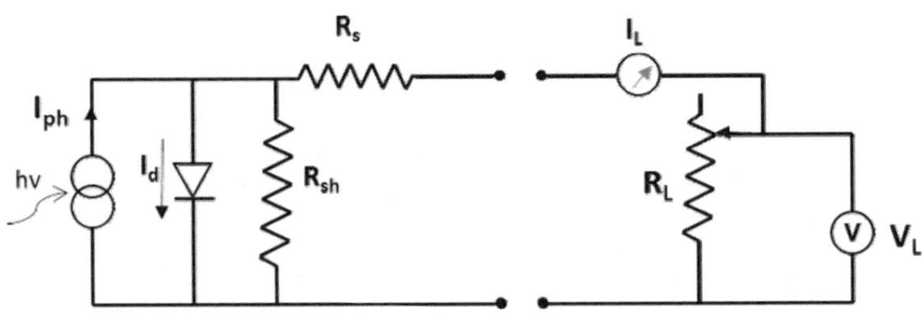

Fig. 4.14. Equivalent circuit diagram of solar cell.

The photo generated current I_{ph} is shown by a current generator. The magnitude of I_{ph} depends on the optical absorption coefficient of solar cell, wavelength and intensity of incident light, width of depletion region, life time and mobility of carriers from both sides of junction.

The polarity of the output photo-voltage V_L is such that the diode is in forward bias condition. The current flowing through diode I_d can be given by

$$I_d = I_0\left[exp\frac{eV_L}{nKT} - 1\right] \tag{4.5}$$

Where, I_o is the reverse saturation current, n is the junction ideality factor, and K is the Boltzmann constant.

The shunt resistance (R_{sh}) can be caused by stacking faults, leakages on the surface edges of cell, by diffusing spikes along the defects like dislocations or grain boundaries or by microcracks consisting of metallic bridges.

The component Rs from equivalent circuit diagram shows the resistance element of solar cell known as series resistance, it can arise from front and back contact resistances, resistance from base region itself, and a sheet resistance of thin diffused/grown layer. The load resistance is represented by R_L and it is desirable to choose its magnitude such that the maximum power extracted from the solar cell.

4.7.2. Output parameters of solar cell

The output parameters of solar cell can be estimated from the current-voltage properties of p-n junction. The current-voltage properties of solar cell can be measured in three different ways such as (a) photovoltaic output with constant illumination (b) photovoltaic output with variable illumination (c) diode forward characteristics without illumination. Among these the most common method used to establish output parameters of solar cell is measurement of current - voltage properties under constant illumination of light of known intensity. In this method the resistive load across the terminals of solar cell is varied from very zero load/ short circuit condition to infinite load/open circuit condition and during this variation a pair of current and voltage across the solar cell is recorded. The I-V characteristics obtained in this manner is called as "photovoltaic output characteristics". The graph of photo-voltaic output characteristics is shown in Fig. 4.15 below, and the circuit diagram of particular setup of measurement is similar to Fig 4.14. On illumination I-V characteristics are obtained in forth quadrant, as the solar cell acts as the generator of energy in this method. Here we can observe that an I-V characteristic of solar cell under illumination is the superposition of I-V characteristics of solar cell diode in dark with addition of photo-generated current.

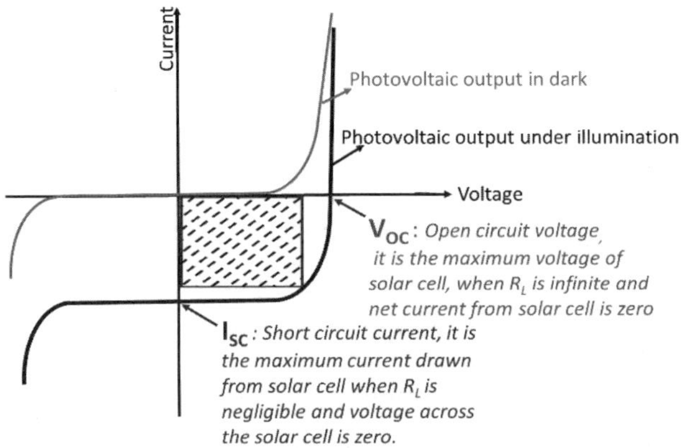

Fig. 4.15. Output characteristics of photovoltaic solar cell.

4.7.2.1. Short circuit current I_{sc}

The current through the solar cell at zero voltage is known as short circuit current (I_{SC}). Also we can say the current through solar cell at short circuited condition. At short circuited condition the load resistance R_L is zero and hence V_L is zero. It is the largest current that can be flown through the solar cell. In short circuit condition the resistive loss of current is very less hence the light generated current and I_{SC} are similar. Thus we can write.

$$I_L = I_{sc} = I_{ph}$$

The value of I_{SC} depends on various factors like (i) Area of solar cell (ii) number of incident photons (iii) spectrum of incident light (iv) collection probability of solar cell (v) Optical properties of solar cell. To remove the dependency of I_{sc} on area of solar cell, it is commonly measured as short circuit current density (J_{sc}). Where the J_{sc} is current per unit area mA/cm^2.

4.7.2.2. Open circuit voltage V_{oc}

It is the maximum voltage across the terminals of solar cell, when current flowing through the solar cell is zero and R_L is infinite. Thus under these condition one can write equation for V_{OC} as

$$V_{oc} = \frac{KT}{e} \ln\left(\frac{I_{sc}}{I_o} + 1\right) \tag{4.6}$$

Considering the conditions that temperature is room temperature and $I_{sc} >> I_o$, we can modify above equation as

$$V_{oc} = 0.0575 \ln\left(\frac{I_{sc}}{I_o}\right) \tag{4.7}$$

From this equation we can understand that V_{oc} is determined by saturation current and photogenerated current. Where the variation in I_{SC} is typically less, hence the key effect is of saturation current. The V_{OC} of solar cell can be increased by minimizing I_o. The saturation current is determined by the recombinations in solar cell. For a monocrystal I_o can be given by

$$I_o = qn_i^2\left(\frac{D_n}{N_a L_n} + \frac{D_p}{N_d L_P}\right) \tag{4.8}$$

Where N_a and N_d are the minority charge carriers from p and n layer respectively. And D_n and D_p are corresponding diffusion coefficients. L_n is electron diffusion length, L_p is the hole diffusion length. The n_i is the equilibrium population of electrons. Thus from equation (4.6), (4.7) and (4.8) it is seen that Voc is determined by the properties of semiconductor by the virtue of its dependence on I_o. Further it can be shown that for a larger band gap material we can get a larger open circuit voltage. For a given band gap, material should have low carrier mobility, low resistivity and high minority carrier life time.

4.7.2.3. Fill factor

The Fill factor of solar cell can be defined as it is the factor representing the correctness/sharpness of the knee of output current voltage curve of solar cell performance. The expression for fill factor for can be written as given below. The maximum value of FF can be 1.

$$FF = \frac{I_{mp} \times V_{LMP}}{I_{sc} V_{oc}} \tag{4.9}$$

4.7.2.4. Efficiency of solar cell

Efficiency of any device is the ratio of power output and power input. Maximum power output is given by

$$P_{\max} = I_m \times V_m \tag{4.10}$$

From the equation of fill factor we can write

$$P_{\max} = I_{sc} \times V_{oc} \times FF \tag{4.11}$$

Thus expression for efficiency of solar cell can be written as

$$\text{Efficiency (n)} = \frac{I_m \times V_m}{P_{in}} \tag{4.12}$$

And

$$\text{Efficiency (n)} = \frac{I_{sc} \times V_{oc} \times FF}{P_{in}} \tag{4.13}$$

Considering the schockley-quissor limit for single p-n junction the maximum efficiency obtained from solar cell can be 43%.

4.7.2.5. Influence of series and shunt resistance

The two principle resistive components of solar cell are (1) series resistance R_s and (2) Shunt resistance R_{sh}. In earlier section we have studied the origin of these resistances. Here we will study how to evaluate R_s and R_{sh}, also we will study its influence on performance of solar cell.

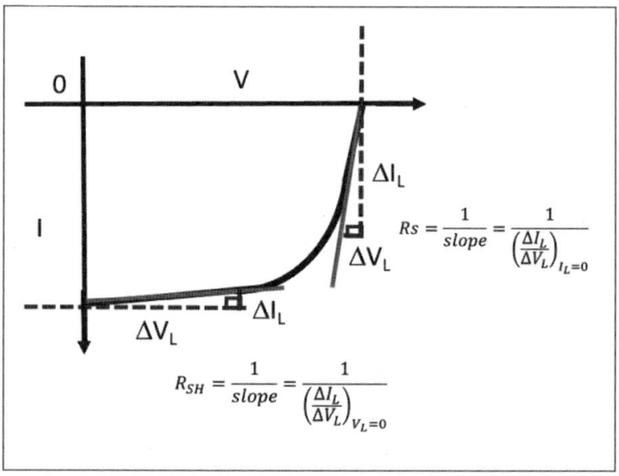

Fig. 4.16. Determination of R_s and R_{sh} values form I-V curve

When the I-V characteristics are measured under constant illumination then R_s can be evaluated by taking slope at $I_{sc} = 0$ and R_{sh} can be evaluated from taking slope at $V_{oc} = 0$ as shown in fig 4.16.

$$R_S = \frac{1}{slope} = \frac{1}{\left(\frac{\Delta I_L}{\Delta V_L}\right)_{I_L=0}} \tag{4.14}$$

$$R_S = \frac{1}{slope} = \frac{1}{\left(\frac{\Delta I_L}{\Delta V_L}\right)_{V_{L=0}}} \tag{4.15}$$

The series resistance mainly composed of contact resistances and bulk resistance of solar cell. The relation between photocurrent and Rs is given by equation

$$I_{sc} = I_o\left\{exp\left[\frac{e}{nkT}(V_{oc} - I_L R_S)\right] - 1\right\} - 1_{ph} \tag{4.16}$$

At open circuit condition $I_L=0$, putting $I_l=0$ the expression becomes similar tó relation for V_{oc}. It reveals that change in R_s does not affect the value of V_{oc}, but it value of Isc strongly depends on the R_s. The Fig. 4.17 given below shows the theoretical I-V characteristics of solar cell at different values of R_s. This image elucidates the relation between R_s, I_{sc} and FF.

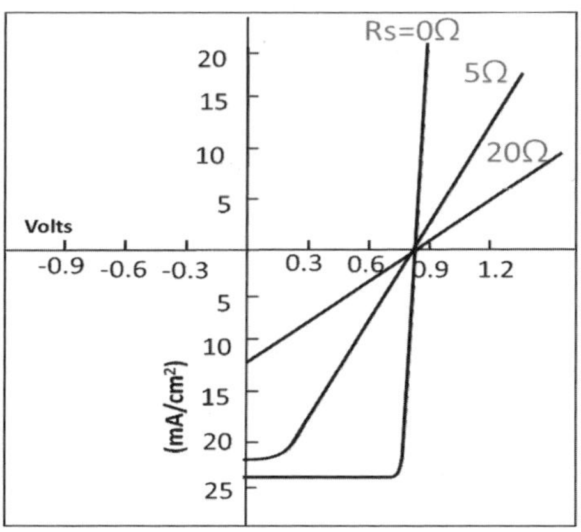

Fig. 4.17. Plausible J-V characteristics of solar cells with different values of R_S.

The influence of R_{sh} on photovoltaic properties of solar cell is shown in Fig. 4.18 below. From the fig we can understand that the change in R_{sh} does not affect the value of I_{sc} but the value of V_{oc} predominantly influenced by the R_{sh}. From the Fig. 4.17 and 4.18 it reveals that the efficiency of solar cell predominantly depends on values of R_s and R_{sh}. For better performance of solar cell the value of R_s should be very less (nearly zero) and R_{sh} should be very high (infinite).

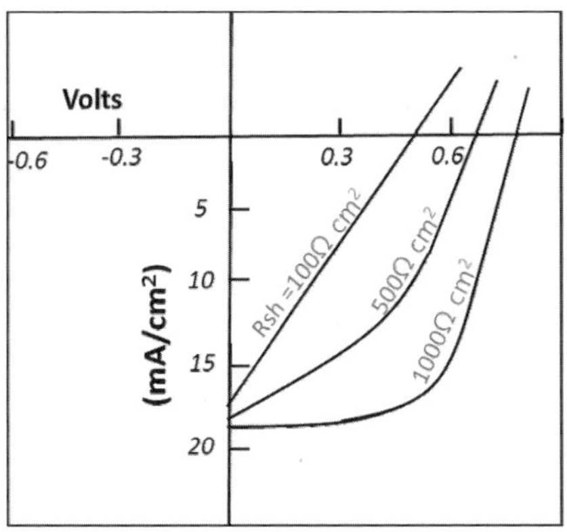

Fig. 4.18. Plausible J-V characteristics of solar cells with different values of R_{sh}.

4.7.2.6. Influence of light intensity

All the solar cell parameters like short circuit current, open circuit voltage, fill factor, efficiency, series and shunt resistance gets influenced by change in light intensity. The maximum impact of light intensity variation observes in short circuit current density;

it increases linearly with increasing light intensity. The current-voltage characteristics of solar cell with different light intensities are shown in fig. 4.19 below. Considering the equation 4.6 it seems that open circuit voltage increases logarithmically with increasing light intensity. Thus increasing short circuit current density and open circuit voltage leads to further enhancement in solar cell efficiency with increasing light intensity.

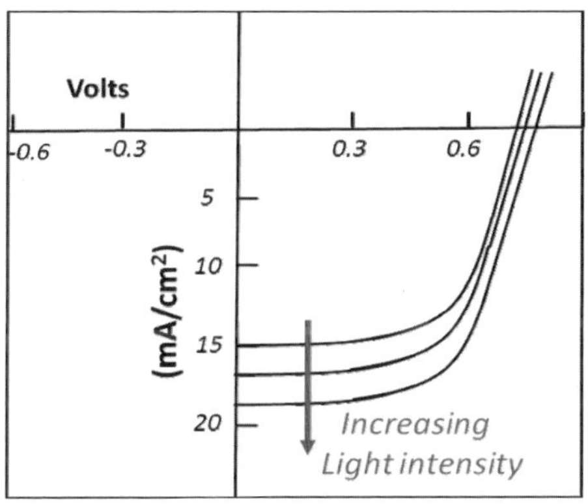

Fig. 4.19. Effect of light intensity variation on the J-V characteristics of p-n junction solar cell.

The incident light intensity is measured in suns, where the one sun is 1kW/m^2. This is equivalent to the standard sun illumination at AM1.5. If 10 kW/m^2 light is illuminated on solar cell then solar cell is said to be operating at 10 Sun. With increasing light intensity the series resistance of solar cell and temperature of cell increases; this tends decrease the solar cell performance. The net effect is the increase in efficiency with increase in light intensity; as the increase in J_{sc} is quite higher than increase in R_s. Moreover the increase in R_s occurs at very high values of suns nearly some hundreds of sun. Hence up to the certain limit of light intensity (usually few hundred suns) solar cell performance increases with intensity.

4.7.2.7. Influence of temperature

Solar cell performance is quite sensitive to the temperature; the most impacted parameter of solar cell is open circuit voltage. The equilibrium population of electrons n_i increases exponentially with increase in temperature; it leads to increase in dark saturation current density.

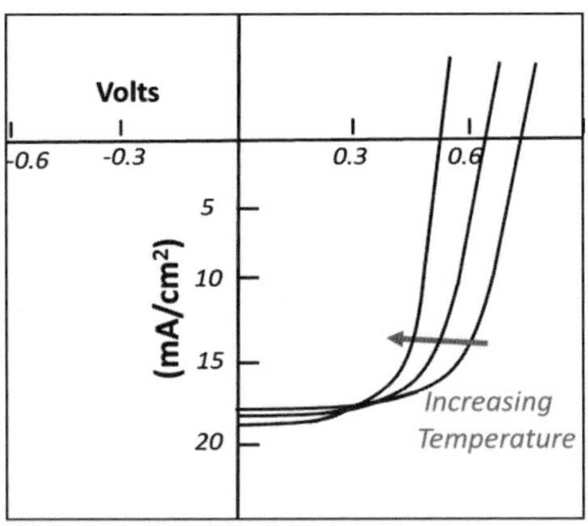

Fig. 4.20. Effect of temperature variation on the J-V characteristics of p-n junction solar cell.

Considering the equation 4.7 and 4.8, it seems that increase in dark saturation current density reduces the open circuit voltage.

The effect of increasing temperature on J-V characteristics of solar cell is shown graphically in Fig. 4.20. From figure we can observe that with increasing temperature there is slight increase in short circuit current density Jsc. This enhancement in Jsc can be attributed to lowered band gap, due to increased temperature. Due to reduction in band gap the low energy photons can also now be absorbed leading to increase in photocurrent. But the loss in open circuit voltage outweighs the rise in short circuit current density; hence the net effect is reduction in efficiency with increase in temperature.

References

1. Richard Bube "Fundamentals of Solar Cells," Academic press, inc. Published by Elsevier Inc. 1983.
2. Chenming Hu, Richard M. White, "Solar Cells From Basics to Advanced Systems," University of California, Berkeley McGraw-Hill Book Company 1983.
3. J. P. McKelvey, Solid State and Semiconductor Physics, Harper and Row, New York (1966), Chapter 16.
4. S.O. Pillai, "Solid State Physics," A New Age International Publishers 2001.
5. Martin A. Green, "Solar Cells: Operating Principles, Technology and system applications," 1982.
6. G.C. Jain and W.B. Berry, Transport properties by solids and solid state energy conversion, Tata Mc-Graw Hill, Bombay 1972.
7. L. Solymar and D. Walsh, Electrical properties of Materials, Oxford University Press, Tokyo 1988.
8. http://pveducation.org/
9. http://ncertbooks.prashanthellina.com/class_12.Physics. Physics PartII/index.html.
10. Jenny Nelson "Physics of solar cells," Imperial college press 2004.

Chapter 5
Thin film solar cells

5.1. Introduction

In previous chapter we have learnt the basic physical properties of semiconductor materials, p-n junction and photovoltaic effect. Among various solar cells, the wafer based crystalline and polycrystalline silicon dominates the market. However to reduce the price and avoiding the costly wafer process thin film solar cells were developed. For the last Six decades various thin film solar cells based on Cu_2S/CdS, a-Si, $CuInGaSe_2$, CdTe, CZTSSe and CTS solar cells were researched as an alternative to expensive silicon solar cells due to large area deposition and less material utilization with high throughput on in-expensive glass substrates. In this chapter we learn basic fabrication of most promising thin film solar cells based on CIGS, CdTe, CZTSSe and CTS. The first two are successful materials for high efficiency thin film solar cells with conversion efficiencies over 20% were achieved. On the other hand the last two materials are emerging as potential replacement for CIGS and CdTe due to earth abundant and less toxic elements. Understanding these abundant materials will pave way for developing low cost thin film solar cells. In this present chapter the material properties and fabrication of these solar cells were discussed.

5.1.1. CIGS thin film solar cell

Cu(InGa)Se_2-based solar cells have been considered as promising of solar cell cost-effective technology compared to silicon solar cells. The main driving force for development of these solar cells is advantages of low cost, high throughput, high rate deposition on large area thin film solar cells and fabrication of monolithically interconnected modules. This will reduce the cost in comparison to bulk crystalline silicon solar cells due to elimination of bulk crystal growth and making wafers.

5.1.2. Crystal structure of $CuInSe_2$

In the 1950's Bell labs in USA grown ternary compounds of chalcopyrites and investigated its structural, electronic and optical properties of these materials. I-III-VI_2 compounds $CuInSe_2$ and $CuGaSe_2$ have the chalcopyrite crystal structure. This structure is analogous to sphalerite (ZnS) crystal structure. Replacing Cu and In or Ga elements on Zn site in ZnS results tetragonal unit cell as shown in Fig. 5.1 with lattice parameters a and c with ration of c/a = 2, however the ratio of c/a is smaller than 2 due to tetragonal distortion arising from different bond lengths of Cu-Se and In-Se and Ga-Se bonds.

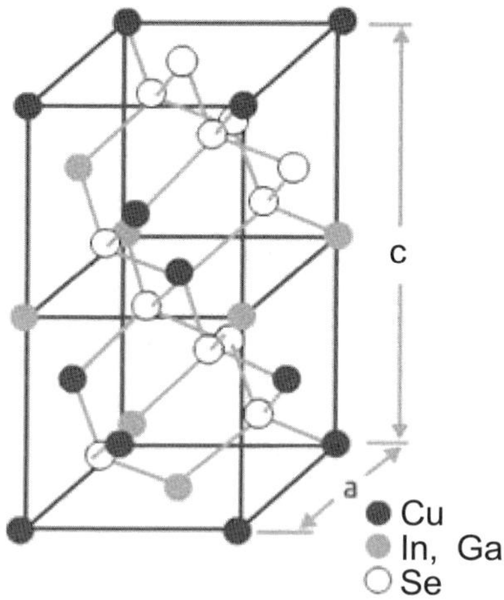

Fig. 5.1. CIGS tetragonal crystal structure.

$CuInSe_2$ phase forms between the binary Cu_2Se and In_2Se_3 tie line in the phase diagram as depicted in Fig. 5.2.

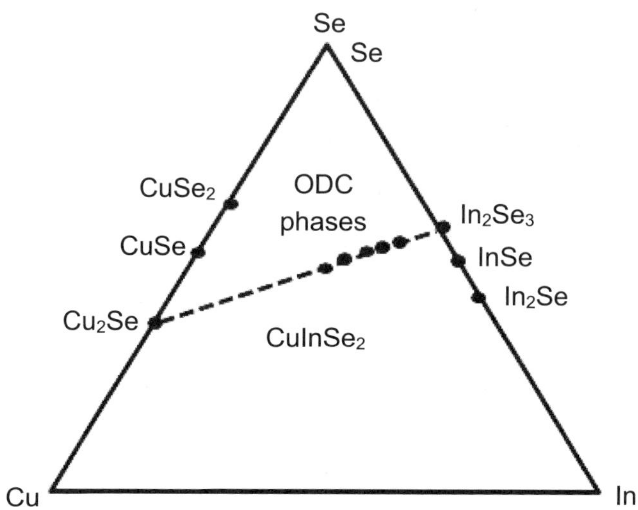

Fig. 5.2. Ternary phase diagram of Cu-In-Se.

The formation of $CuInSe_2$ phase region is very narrow which contains less than 25% Cu. Binary and ordered vacancy compounds forms due to off-stoichiometric composition. In the diagram it can be seen that the chalcopyrite $CuInSe_2$ α, high-temperature δ phase with the sphalerite structure and an ordered vacancy compound (ODC) β phase.

5.1.3. Optical properties

$CuInSe_2$ is a direct band gap with very high absorption coefficient α. One of the basic requirement for thin film solar cells is absorption coefficient, higher the 'α' lesser the thickness required for incident solar absorption. The fundamental absorption edge is described by the following relation.

$$\alpha = A(h\nu - E_g)^2/E \tag{5.1}$$

The band gap of $CuInSe_2$ is 1.04 eV. The optical band gap of this material can be tuned by adding Gallium to this material. The resultant material is known as $CuIn_{1-x}Ga_xSe_2$. The variation of band gap with composition (x) is given below according to the equation [1].

$$E_g = 1.010 + 0.626x - 0.167x(1-x) \tag{5.2}$$

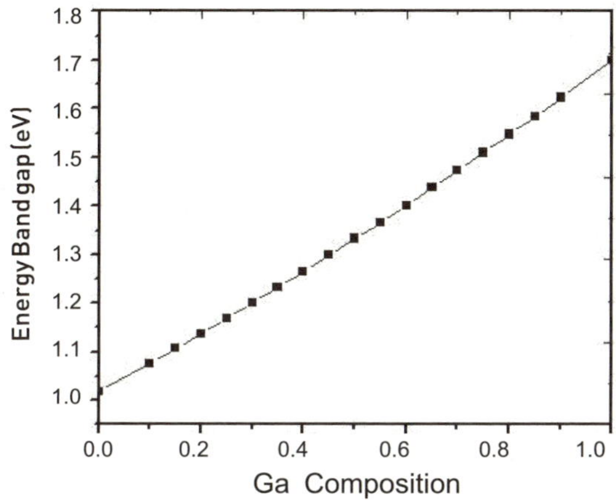

Fig. 5.3. Schematic diagram of a thin film CIGS solar cell.

Figure 5.3 shows the variation of band gap of CIGS as function of 'Ga' content in the films. The band gap of this material can be tuned from 1.04 to 1.68 eV to match the solar spectrum for maximum conversion efficiency.

5.1.4. Electrical properties

$CuInSe_2$ (CISe) and $CuGaSe_2$ (CGSe) exhibits p-type conductivity. Generally, in elemental semiconductors conductivity type is due to doping of impurities of trivalent Boron, Aluminum, Gallium and pentavalent Phosphorus and Arsenic. The conductivity type of CISe strongly depends on composition, Cu rich CISe always shows p-type, whereas Indium rich compound shows p or n-type conductivity depending annealing atmosphere. In case of CGSe always exhibits p-type conductivity for any composition. In chalcopyrite compounds conductivity originates from intrinsic defects. Numerous intrinsic defects such as copper vacancies V_{Cu}, anti-sites Cu_{In}, Cu_{Ga}, Ga_{Cu}, In_{Cu}, V_{In}, V_{Ga}, and V_{Se} forms in the compounds. V_{Cu} has low formation energy and is responsible for p-type behavior in CISe and CGSe materials [2]. The device quality $CuInGaSe_2$ thin film absorbers have typically carrier concentration values about $10^{16}/cm^3$.

5.1.5. CIGS thin film solar cells

First thin film solar cells were fabricated with $CuInSe_2/CdS$ with evaporation of evaporation of $CuInSe_2$ powder. However, with demonstration of 9.4% solar cells by Boeing from co-evaporation process has attracted wide attention due to potential application of this material. From then with improved understanding in the growth of CISe and CGSe high efficiency solar cells were fabricated with empirical approach with various modification of processes, materials and the band gap of the absorber is increased with alloying gallium from 1.04 to 1.2 eV. And replacing a very thick (CdZn)S layer with a thin CdS layer and window layer consists of conducting ZnO to reduce the losses at short wavelength. Use of soda-lime glass substrates the performances were improved due sodium doping from the glass substrate.

The microstructure of the CIGSe films is markedly improved due to understanding of crystallization of the absorber with presence of liquid phase $Cu_{2-x}Se$ which assists large grain CIGS absorbers with high electronic quality. These steps implemented for last three decades to enhance the performance of the CIGS thin film solar cells from 9.4 to 21.7% using various modifications [3]. The schematic solar cell is shown in Fig. 5.4. The improved performances were achieved with optimization of each layer with precise control of each process steps lead to higher conversion efficiencies.

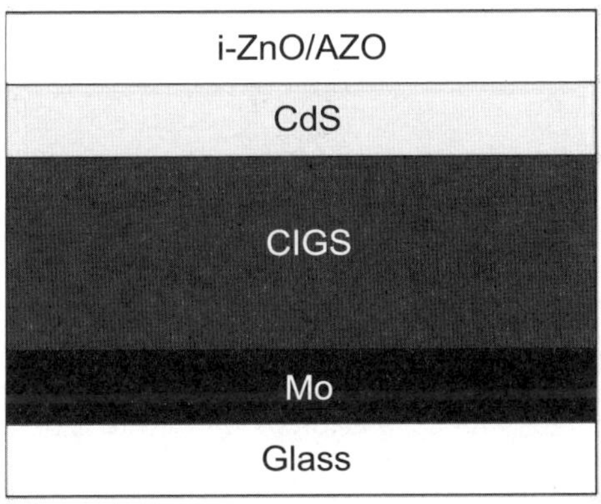

Fig. 5.4. Schematic diagram of a thin film CIGS solar cell.

Substrates

Various substrates such as glass, flexible steel or polymide were used for fabrication of CIGS solar cells. Soda-lime glass is commonly used as substrate for fabrication of high efficiency solar cells. The advantage of using Soda-lime glass is sodium Na incorporation into the growing absorber film which yielded the highest efficiencies in laboratory scale devices.

On the other hand barrier layers such as Si_3N_4 and Al_2O_3 were used to prevent the unwanted iron Fe impurities in the absorber layer which reduces the efficiency due to shunting. To supply Na into the flexible metal substrate a sodium compound NaF is generally deposited onto Mo. In case of polymer also Na is supplied externally by means of physical or chemical methods. Both flexible solar cells are light weight compared to rigid glass substrate solar cells. The flexible solar cells are suitable for wide range of building integrated photovoltaic solar cells and mobile applications.

Back contact

Molybdenum is typically used for back contact in CIGSSe thin film solar cells. Molybdenum meets the criteria as ohmic contact and chemically inert during the absorber formation at temperatures above 500°C. Various other transition metals are also used as back contact in thin film solar cells. However, they are not as efficient as Mo due to reaction with CIGSSe and diffusion into the absorber. DC sputtering is generally used for deposition of Mo back contact. The properties for back contact are about 1 μm thick film with contact resistivity value about 0.3 Ωcm. Generally, the sputtered Mo films have a sheet resistance of 0.4-0.6 Ω/square for 1 μm thick films.

CIGS absorber

In CIGS thin film solar cells CIGS absorber is deposited using various deposition tcchniqucs such as co-evaporation, Sputtering, non-vacuum process and sulfo-selenization of CuGa/In metallic precursor layers in selenium, and sulfur atmosphere. Here we describe two widely used methods co-evaporation and selenization of metallic precursors. In co-evaporation process Cu, In, Ga and Se were evaporated onto substrates from the effusion cells as shown in the Fig. 5.5.

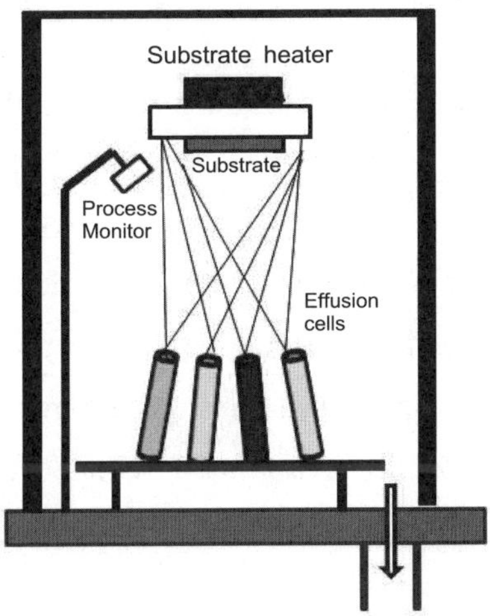

Fig. 5.5. Schematic Co-evaporation system for CIGS deposition.

The figure shows widely used three stage processes in the laboratory scale for high efficiency CIGSe thin film solar cells with over 20% [3].

In the first step In, Ga and Se is evaporated to form $(InGa)_2Se_3$ precursor layer at low temperatures at 300-350 °C after this step the effusion cells of In and Ga is switched off and Cu effusion cell is on, and temperature is ramped to high temperature about 550-600 °C.

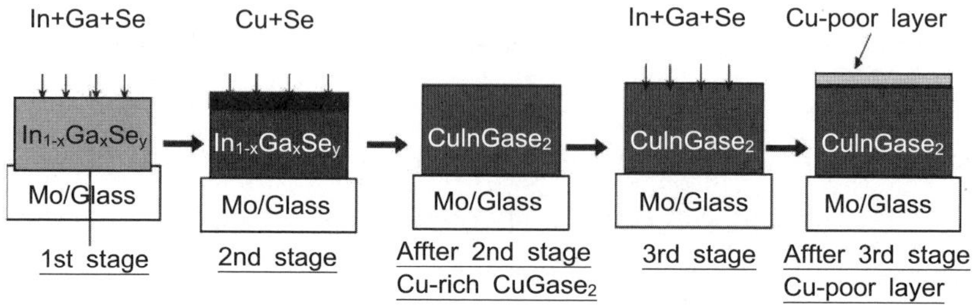

Fig. 5.6. Schematic diagram of three-stage growth process.

During this stage the film undergoes crystallization due to presence of liquid $Cu_{2-x}Se$ at 523 °C and assists formation of large grain CIGS films due to liquid assisted grain growth from flux of this phase. After second stage termination, the film is Cu rich which is not suitable for high efficiency due to presence of highly conducting phase $Cu_{2-x}Se$ at the surface. To remove this phase in third stage In and Ga are evaporated with Se to complete the growth process to achieve slightly Cu-poor composition of CIGS absorber layers.

The efficiency CIGS solar cells depend on the Ga content in the $CuIn_{1-x}Ga_xSe_2$. The control of Ga/In+Ga ratio in the absorbers is very important to achieve high efficiency CIGS solar cells [4]. The efficiency decreases with increase in Ga content above 0.3 in the absorbers. Hence, it is essential to control the ratio of Ga/Ga+In in the sample. This can be achieved by controlling the deposition rates of gallium and indium sources. The co-evaporation of Cu, In, Ga and Se can be done in many ways such as co-evaporation of elements simultaneously, in two - step and linear evaporation on large scale. In order to make large area CIGS modules linear co-evaporation is developed.

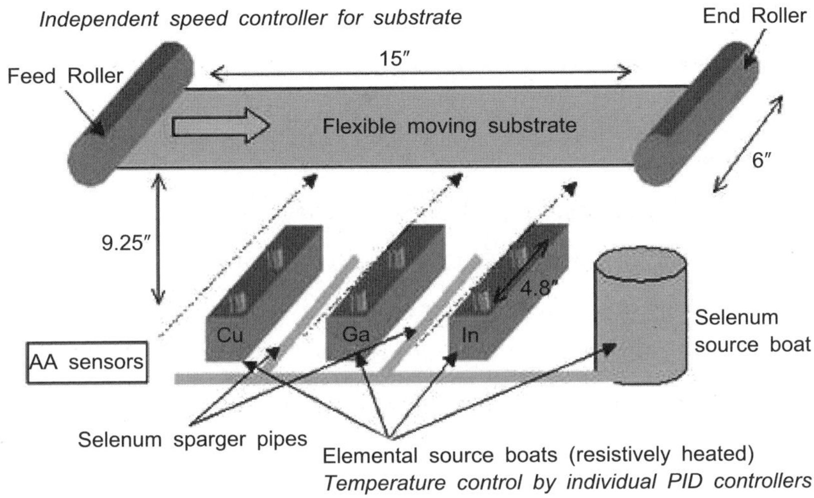

Fig. 5.7. The schematic diagram linear co-evaporation of CIGS absorbers [5].

Two step process

In another different way of making CIGS absorbers is two-step process. In this process the metallic layers CuGa/In was sputter deposited on to Mo coated glass substrates

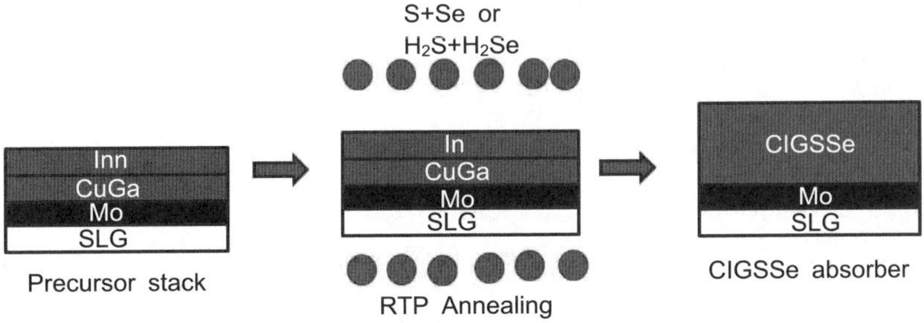

Fig. 5.8. The schematic diagram of two step-process for growth of CIGSSe absorbers.

During metal layer deposition the layers diffuse and form intermetallic CuGa, Cu In alloys. In the second step these precursors were annealed in selenium or sulfur containing atmosphere to convert into CIGS Se absorbers. The schematic process of two-step for absorber growth is given in Fig. 5.8. This process is easily scalable for large area due DC sputtering of metal layers at room temperature and thermal annealing in the furnaces. However, the quality of the absorber grown in this method is inferior to the absorbers grown with three stage process. CIGS absorbers grown by selenization there is accumulation of Ga at the bottom of the film due to difference in reaction of indium and gallium so the films have some phase segregation which leads to grading in the absorbers. So the efficiencies obtained with these absorbers are lower compared to CIGS cells fabricated with three stage evaporation method. With continuous understanding the grading and controlling the time temperature profiles of annealing lead to higher conversion efficiencies over 22 % achieved using this process [6].

CdS

In CIGS solar cell CdS is typically used as a n-type partner in p-n junction. CdS is generally deposited by chemical bath deposition using metal salts. The role of buffer layers is to prevent the shunting paths and protect the absorber from damage during the sputtering of highly conducting transparent oxide (TCO). Typically a 50 nm CdS layer is deposited at 80 °C for 10-20 min on to CIGS absorbers. The reaction of precursor solution leads formation of layer by layer deposition of buffer layers following chemical reaction.

$$Cd(NH_3)_4^{2+} + SC(NH_2)_2 + 2OH^- — CdS+H_2NCN+4NH_3+2H_2O$$

The thickness of the CdS should be thin so as to reduce the losses at blue region of the solar cells. CdS has a band gap of 2.41 eV replacing with large band gap materials has advantages due to less absorption in the blue wavelength region. Various cadmium free buffer layers such as ZnS, ZnO, ZnMgO, ZnSnO and $ZnIn_xSe_y$ were tried as buffer layers in CIGS thin films solar cells. Among these ZnS and ZnSnO buffer layers has been used to fabricate high efficiency CIGS solar cells. ZnS buffer is deposited by chemical bath deposition and ZnSnO by Atomic layer deposition [6,7].

Buffer layer

Typically, an intrinsic zinc oxide is used as buffer layer between n-type CdS and highly transparent conducting oxide top contact. This layer is deposited by RF sputtering using an oxide sputtering target in oxygen and argon atmosphere. This layer has thickness about 50 nm and resistivity about 50-100 Ωcm. It has thought that this layer reduces losses due shunting paths arising from the non-uniformity of the absorbers and CdS layer. And it also protects the solar cells during TCO deposition.

Top contact

One of the important criteria for top contact in CIGS solar cells is highly transparent to reduce the absorption and must have high conductivity to reduce electrical losses. Typically, Indium tin oxide (ITO) and Al doped zinc oxide (AZO) are used as transparent contacts in CIGS solar cells. Among the two AZO is widely used as top contact due to low cost. The AZO is deposited by RF sputtering using sintered oxide targets. About 500 nm thick layer with a sheet resistance 20-30 Ω/square is used for small scale devices and sheet resistance of < 10 Ω/square was used for large area modules to reduce the electrical losses due to higher currents. By following the above procedure one can fabricate high efficiency CIGS thin film solar cells.

The same procedure is applied for fabrication of large area CIGS thin film solar cell modules. The process sequence used for module manufacturing using two-step process is shown in Fig. 5.9. The incoming glass is cleaned and dried after this Mo back contact is deposited on to glass substrate and the Mo contact is scribed by laser P1.

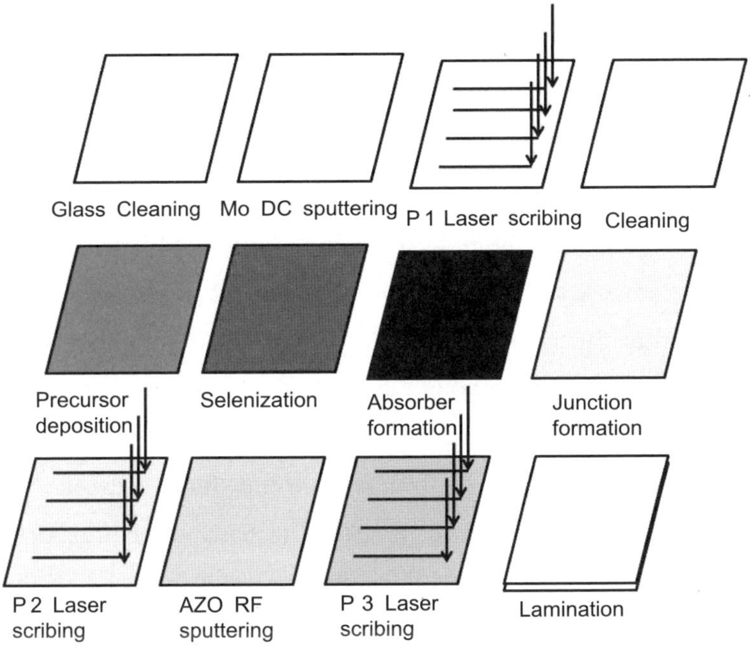

Fig. 5.9. The schematic diagram process flow for CIGS modules.

After cleaning the dust, the precursors are deposited in case of two-step or co-evaporation of absorber is performed at this stage. As deposited precursor is converted into absorber by annealing in selenium/sulfur atmosphere. The buffer layer is deposited to form p-n junction followed by isolation with laser scribe P2. Finally the window layer and TCO are deposited by RF sputtering and final cell isolation is done by scribing with laser P3. At the end the module is laminated by placing a front glass and current voltage characteristics are performed.

5.2. CdTe thin film solar cells

Another material has attracted must attention for thin film solar cells was CdTe. It is a simple binary II-VI compound with a direct band gap of 1.4 eV which is optimum for high efficiency solar cells and has high absorption coefficient over 10^5 cm^{-1}. Extensive work was carried out on this material in single crystalline and thin film form for last five decades. Due to simplicity of this material system with suitable electro-optical properties, CdTe has considered as potential material for solar cells. Numerous methods were developed to deposit polycrystalline thin films and solar cells. In the following sections we describe the basic structural, optical and electrical properties of CdTe and various deposition methods used to depasit CdTe thin films and development of CdTe thin film solar cells.

5.2.1. Structural properties

Binary CdTe crystallizes in face centered zinc blend structure with space group F-43 m and with lattice parameter 6.481 Å. Among binary compounds CdTe has high iconicity. The atomic radius of the Cd^{2+}/Te^{2-} favors tetrahedral coordination with bond length 2.806 Å between Cd and Te. Fig. 5.10 shows the zinc blend crystal structure of CdTe.

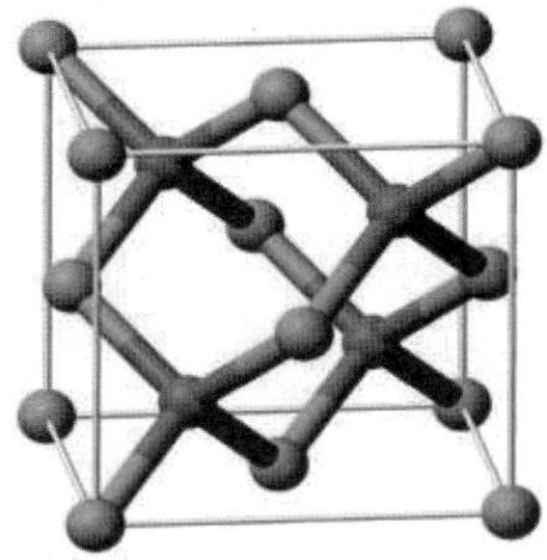

Fig. 5.10. shows the crystal zinc blend structure of CdTe.

5.2.2. Optical properties

CdTe is a direct band gap semiconductor with a band gap of 1.4 eV that means that the conduction band minimum (CBM) and valence band minimum occurs at same energy level in the energy-momentum (E-k) curves. In CdTe valence band is made of 5p levels of Te and conduction band is formed from the 5s level of Cd.

5.2.3. Defects and electrical properties

The electronic and optical properties of CdTe strongly depend on composition. Any deviation from the composition results in formation of intrinsic defects. The possible defects that forms in CdTe are Cd vacancy V_{Cd}, cadmium on tellurium CdTe forms shallow acceptor states. On other hand tellurium interstitial Tei forms deep states which can act as recombination centers in the material [1].

5.2.4. Solar cell device and fabrication

Compared to CIGS thin film solar cells which is fabricated in the substrate configuration. The CdTe thin film solar cells are fabricated in the superstrate configuration TCO/CdS/CdTe as given in Fig. 5.11.

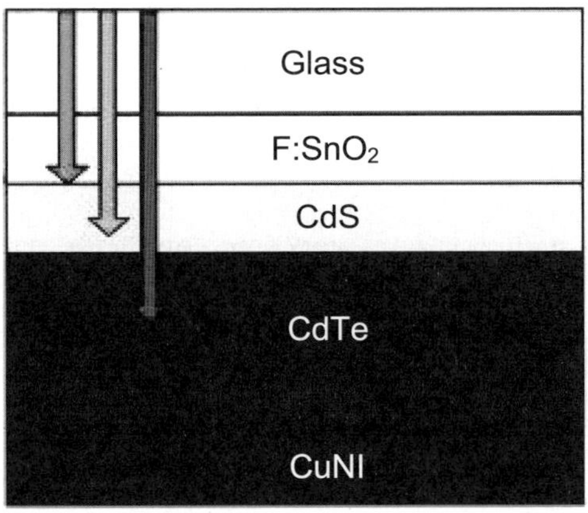

Fig. 5.11. Schematic diagram of the CdTe thin film solar cells.

The layers were deposited sequentially with start of transparent conducting oxide, CdS and CdTe. Solar cells fabricated on substrate configuration exhibited poor performance due various factors such as low CdS/CdTe junction quality and stability of back contact, Cu diffusion into CdTe and high barrier at the back contact. Hence, superstrate configuration is most widely used for CdTe solar cells.

Top contact: The top contact is a transparent conductor, typically used TCO for CdTe is fluorine doped tin oxide SnO_2:F is deposited onto glass substrates. The choice of SnO_2:F as TCO compared to ITO, AZO due to chemical inert and having high transparent, conductivity and low cost.

CdS deposition: The n-type CdS is deposited onto TCO by various methods such as chemical bath deposition, vacuum evaporation and vapor transport method. As mentioned previously vapor transport method is widely used for large area deposition. The deposited films are thin enough to allow more light with less absorption at blue region of the spectrum.

CdTe absorber: Deposition methods

Various chemical and physical techniques methods were used to deposit polycrystalline cadmium telluride thin films. Among these techniques, eight methods were used successfully to grow CdTe thin films. They are screen printing, spray pyrolysis, metal organic chemical vapor deposition (MOCVD), electro deposition, physical vapor deposition (PVD), close spaced sublimation (CSS), vapor transport deposition (VTD) and sputtering. In the following section three physical methods were described due to their wide use in growing polycrystalline CdTe thin film solar cells.

Physical vapor deposition

CdTe films were grown by evaporating Cd and Te powders or sublimation of pre synthesized CdTe powders. Comparing the vapor pressures of elemental Cd and Te, the vapor pressure is very low which facilitates growing CdTe films over wide range of substrate temperatures. The vacuum evaporation can be carried out using the system used for CIGS thin film solar cells. One can control fluxes of Cd and Te evaporating from effusion cells. Grain sizes over 1 μm were obtained with (111) orientation for evaporated CdTe films [1].

Closed spaced sublimation (CSS)

CdTe films can be deposited below 350°C by evaporation increasing substrate temperature above 400°C leads re-evaporation of CdTe from the substrate which results poor morphology of the films. To overcome this closed spaced sublimation (CSS) was used to deposit CdTe thin films. In CSS the source and substrate distance is very small. The deposition is done in the presence of inert gas atmosphere consists of N_2, He, Ar etc. Using CSS CdTe films grown on substrate temperatures above 550°C. In order to obtain absorbers for solar cell application the CdTe films always grown in the presence of oxygen O_2. The film grown with this technique showed random orientation with micron sized grains. This process is widely investigated for production of CdTe solar cells due to high quality of the CdTe absorbers [1].

Vapor transport deposition

In vapor transport deposition (VTD) the CdTe vapors were created by heating the sources to a high temperature and using a carrier gas to carry these vapors onto moving substrates kept at high temperatures. The carrier gases are N_2, He, Ne, Ar and O_2. Compared to CSS the VTD is high rate and high throughput process. The films were deposited onto large area substrates in moving condition. The microstructure grown in this method is randomly oriented and grain sizes are comparable to the film thickness. This method is widely used in production of CdTe solar cell modules presently [8].

However, as deposited CdTe is in low quality for fabrication of solar cells due to formation of defects, Generally, CdTe films were post heat treated in the presence of $CdCl_2$ to improve the properties of the films. The CdTe films were exposed to $CdCl_2$ and oxygen environment at 350 - 450 °C for 15 to 30 min. The heat treatment has profound influence on the microstructure of the CdTe films. The heat treatment promotes grain growth, recrystallization, and grain orientation due to flux effect of $CdCl_2$ During the heat treatment the reaction between CdS and CdTe occur and inter mixing the two compounds results new compound $CdTe_{1-x}S_x$ with partial substitution of sulfur for tellurium. The formation of this compound has no effect on the efficiency of the devices due to similar crystal structure and having nearly band gap of 1.39 eV as the parent compound CdTe.

Finally top contact is deposited on the CdTe. In order to reduce the barrier at the back the surface of CdTe is treated to make slightly Te rich by chemical etching. The Cu is deposited onto the surface the cupper reacts with tellurium and forms Cu_2Te.

Thin film solar cells fabricated from the above method showed conversion efficiencies over 20%. Presently, CdTe is fabricated using vapor transport deposition on large area modules which are commercially available. The thin films CIGS and CdTe were characterized using various characterization methods and its photovoltaic performances were measured using illuminated current-voltage, dark I-V and spectral responses of the devices in order to understand the behavior and further improvement of the performances

5.3. Kesterite-based thin film solar cells

Recently, Kesterite $Cu_2ZnSn(S,Se)_4$ (CZTSSe) semiconductor compound has received tremendous attention has a promising candidate for future low-cost, high efficiency thin film solar cells (TFSCs). This compound consists of solely earth-abundant, non-toxic and low-cost elements, which further proves its significance for both low-cost and eco-friendly TFSCs. In addition, CZTSSe exhibits a high absorption coefficient over 10^4 cm^{-1} and optical band gap energy in the range of 1.0 - 1.5 eV, which is an optimum band gap energy for harvesting maximum solar energy. It was recognized to be a suitable candidate for TFSCs as reported by Ito et. al. [9] for the first time in 1988. A first heterojunction diode exhibited an open circuit voltage (V_{oc}) of 165 mV. Further, the effectiveness of this compound by fabricating TFSCs was proved by Katagiri et al. in 1996. The first TFSCs yielded efficiency of 0.66% [10]. The following section in this chapter will deal with the structural, optical and electronic properties of CZTSSe compound. Further, an overview of state of the art efficiencies will be given and the major challenges associated with these kesterite-based TFSCs will be discussed in this section.

5.3.1. Structural properties

The bulk CZTSSe crystallizes mainly into two different crystal structures such as kesterite and stannite [10]. Kesterite structure which is a derivative of the chalcopyrite (CIGS) structure that is itself a derivative of the zincblende or sphalerite (ZnS) structure. Starting with the zinc-blende (group II-VI), the tetragonal chalcopyrite structure can be derived by the cross-substitution and maintain the octet rule. Zn in the zinc blende structure is substituted by Cu (I) and In (III) and by doubling the unit cell. Further the substitution of In (III) partly by Zn (II) and partly by Sn (IV) results into the kesterite structure [12]. This can be better understood from Fig. 5.12. The kesterite structure has the space group $I\bar{4}$, whereas stannite structure has $I\bar{4}m$ derived from the ternary chalcopyrite structure [13]. Both the structures are quite similar to each other, however the difference between kesterite CZTS and stannite CZTS crystals lies in a different order of the cation sub-lattice. In the kesterite structure, one of the Cu atoms occupies the 2a position; Zn and the remaining Cu are ordered at 2d and 2c with Madelung potentials of 15.04 V, 21.88 V and 15.21 V, respectively [14]. In the stannite structure both Cu atoms occupy the 4d position and Zn is at 2a and their corresponding

Madelung potentials are 15.30 V and 21.62 V. Sn is located at 2b in both the structures The Madelung potentials are 35.04 V for kesterite and 34.95 V for stannite [14]. The difference between kesterite CZTS and stannite CZTS crystals lies in a different order of the cation sub-lattice. Although there was an initial debate on whether CZTSSe crystallizes into kesterite or stannite structure, later it became accepted that it crystallizes into the kesterite structure, which is more favorable energetically [7].

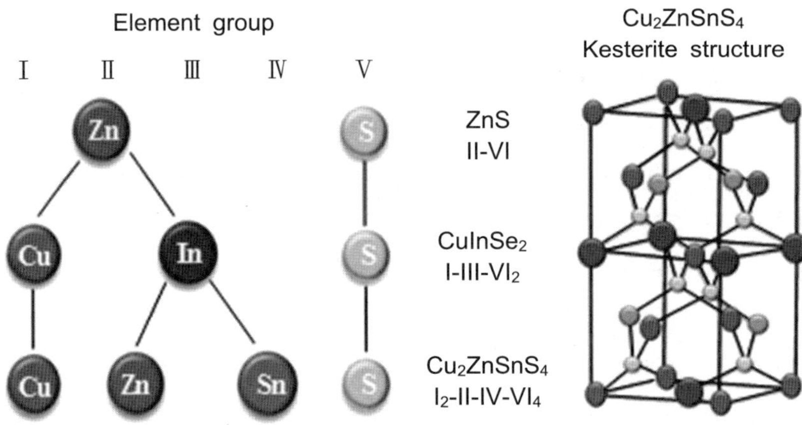

Fig. 5.12. Schematic representation of the origin of CZTS and its kesterite structure.

The kesterite structure which is a derivative of chalcopyrite that is itself a derivative of the zincblende or sphalerite (ZnS) structure. Hence as a super-lattice of kesterite, zinc-blende creates some difficulties in structural crystal analysis [11]. It is very difficult to easily differentiate ZnS(e), $Cu_2SnS(e)_3$ and $Cu_2ZnSnS(e)_4$ phases due to their similar lattice constant. Thus, the presence of ZnS(e) and $Cu_2SnS(e)_3$ along with kesterite $Cu_2ZnSnS(e)_4$ cannot be excluded by XRD alone. The kesterite is a crystal structure, however it is used as a general term to refer CZTS, CZTSe and CZTSSe in the research community and it will be used as it is throughout this section.

5.3.2. Optoelectronic properties

Kesterite compound exhibits has absorption coefficient over 10^4 cm^{-1} and the optical band gap in the range of 1.0 – 1.5 eV, which depends on the anion composition [14, 15, 17]. The band gap energy of CZTSSe can also be tuned by alloying or though the cation substitutions, for example Cd in place of Zn and Ge or Si in place of Sn or alloying with Li [18, 19]. Alloying (impurity doping) of different elements creates an extra energy level, which steers lowering the band gap that leads to the multiexciton generation and efficient charge carrier separation promoting better CZTSSe based PV applications [14].

Kesterite is a p-type semiconductor, which is due to the low, often negative formation energy of Cu_{Zn} antisite defects, i.e. the formation of p-type defects. The main p-type defects in CZTSSe are accepted to be copper vacancy (V_{cu}) and Cu_{Zn} antisite defects [21-22]. The later one produces a deep donor level in the band gap and causes significant decrease in the band gap energy, which sheds light on why CZTSSe should be Cu-poor and Zn-rich. More interestingly, this Zn-rich composition may cause the Zn_{Cu} point defects in CZTSSe, which is actually an n-type defect. Nonetheless, CZTSSe exhibits p-type conductivity with the reported hole mobilities in between 1.6 - 40 $cm^2/(v \cdot s)$ [22, 23]. The conductivity has been found in the range of 10^0 - 10^1 Ωcm, which is dependent on both cation and anion ratios. The carrier concentration is in the range of 10^{15} - 10^{18} cm^{-3}, which increases with decreasing anion contents. The resistivity has been found in the range of 0.13-0.74 Ωcm [23-26].

5.3.3. Defects and doping

The chalcopyrites and kesterites are doped by intrinsic defects. It is of immense importance to have the knowledge of these intrinsic defects in chalcopyrites and kesterites to further improve the efficiency of these devices. CZTSSe has a wider range of possible defects such as antisites, vacancies and interstitials depending upon on its growth conditions and variations from stoichiometry [20-21]. These defects can be located shallow or deep in the band gap, and their concentration depends on their own formation energy. The dominant p-type acceptor in kesterite is the CuZn antisite defect. The formation energy for this defect is the lowest and its located at 0.10 eV above the valence band maximum (VBM), whereas the V_{Cu} is at 0.02 eV above the VBM [20].

The Cu-poor and Zn-rich composition is required for high efficiency CZTSSe devices. This composition allows to the increase of shallow V_{cu} in spite of Cu_{Zn} antisite as well as it suppresses the detrimental $Cu_2S(e)$ phase and deep level defects which acts as recombination centers [25]. It assumed that the formation of electrically neutral defect pairs or compensating defect pairs plays an important role because they can passivate deep level defects in the band gap and thus improves the solar cell efficiency [21].

5.3.4. State of the art efficiencies

IBM has reported a world record efficiency of 12.7% for CZTSSe thin film solar cells, which was fabricated using pure solution processing employing hydrazine as a solvent [27]. Hydrazine dissolves the several metal chalcogenides via "dimensional reduction" Using pure solution approach using hydrazine as a solvent in which Cu and Sn chalcogenides (S or Se) and Zn salt are dissolved. By employing a double buffer layer i.e. In_2S_3 and CdS together, a record power conversion efficiency of 12.7% has been reported. However, the drawback of this approach is the use of highly toxic and flammable solvents, which restricts the use of this approach for large area production. So far, most of kesterite TFSCs have been fabricated using two-step deposition process, i.e. room or low-temperature deposition of precursor film and further the annealing of precursor film in chalcogenide containing atmosphere at above 500 °C.

The following table shows the state of the art efficiencies reported for CZTSSe-based TFSCs by laboratories/companies through world-wide.

Table 4.1. The state of the art efficiencies for CZTSSe-based TFSCS reported by laboratories world-wide

Material	Method	Efficiency	Institute	Ref.
CZTSSe	Solution	12.7	IBM	27
	PVD	10.8	Solar Frontier	28
	Co-sputtering	10.2	Nankai Univ./ Queensland Univ.	29
	Electro-deposition	9.9	Korea Institute of Science and Technology, Korea	30
	Evaporation	9.15	NREL	31
	Nanoparticles	9.0	Purdue Univ.	32

5.3.5. Challenges and future prospects

Two types of challenges are found in kesterite i.e. chemical and electrical in nature. As a multi-element system and due to relatively very small singe phase region, the secondary phases are formed easily in kesterite compounds and thus, make it very complex to grow single phase material. The phase diagram of Cu-Zn-Sn-S has been studied by Olekseyuk et al. [33].

Fig. 5.13 shows the pseudo-ternary phase diagram of Cu_2S-ZnS-SnS. According to this phase diagram, binary sulfides and ternary Cu-Sn-S phases may form in addition to CZTS. It seems that the single phase region of kesterite is very small. Although, even a very less is known about the phase formation of pentanary Cu-Zn-Sn-S-Se system, there is evidence to suggest

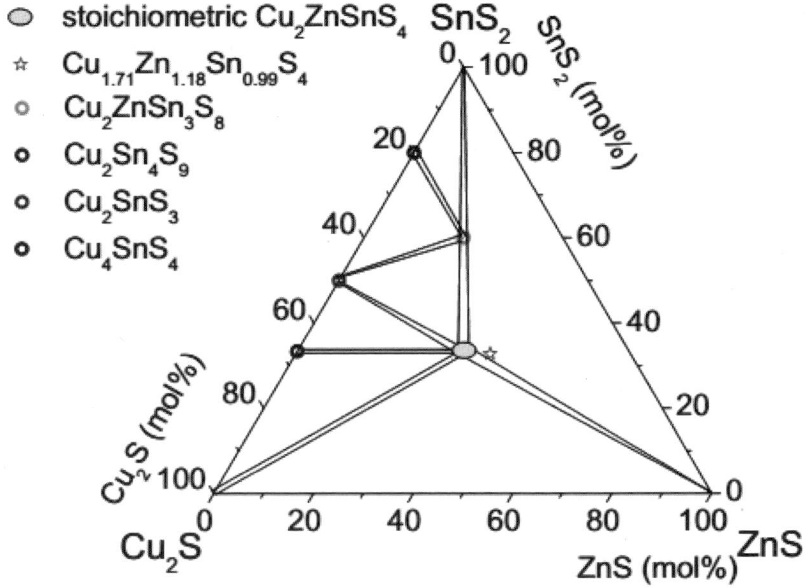

Fig. 5.13. Cu_2S - ZnS - SnS_2 quasi-ternary system adapted from reference [35], showing the excepted secondary phases present in CZTS compound.

that sulfur-containing binary phases would form preferentially over selenium-containing binaries due to lower free energy formation [34]. The presence of secondary phases is undesirable in CZTSSe-based TFSCs as it may act as recombination enters for charge carrier by creating an additional interface in the absorber.

In addition to this narrow region of single phase formation, kesterite suffers from thermal instability. The decomposition starts at temperatures above 400 °C yielding binary sulfides. This frequently observed decomposition reaction at back contact due to instable kesterite/Mo interface was explained by Scragg et al. [36] Decomposition at the back interface due to the formation of MoS_2 and consecutive formation of secondary phases (SnS, Cu_2S, Cu_2SnS_3) could enhance recombination [29]. Further, the kesterite is challenged by several electronically activated defects such as V_{oc} deficit. This V_{oc} deficit could be explained by a short minority carrier lifetime of 3.1 ns, interface recombination or band tailing due to electrostatic potential fluctuations.

To further improve the device performance of kesterite based TFSCs, more detailed understanding of of CZTSSe phase diagram, relation between device characteristics with processing conditions and the formation secondary phases as well as well control over the formation of single kesterite phase without any secondary phase. The major hurdle in kesterite-based TFSCs is the V_{oc} deficit. This could be overcome by deeply studying the formation and impact of bulk, interface and grain boundaries defects, which would solve the problem of V_{oc} deficit in kesterite-based TFSCs. Further, fabrication process for kesterite absorber formation as well as the chemical composition of kesterite absorber should be well-controlled within the narrow phase stability window to avoid the formation of secondary phases under the Cu-poor and Zn-rich conditions. Lastly and most importantly, the interface recombination should be reduced which requires an ideal spike-like conformation of the CBO at the buffer layer/kesterite absorber interface. This ideal spike-like conformation can be achieved by precise controlling the chalcogen space distribution in the absorber as well by adjusting the band structure of the buffer layer or by creating an n-type inversion layer on the surface of the p-type kesterite absorber [37]. Thus, by deeply studying these major hurdles with better understanding could lead to a significant improvement in the device performance of kesterite-based TFSCs.

5.4. Cu_2SnS_3 thin film solar cells

5.4.1. Introduction

CTS (Cu_2SnS_3), a ternary semiconductor compound is an emerging potential candidate for solar cell application. CTS is a P-type semiconductor having high absorption coefficient (10^5 cm^{-1}) and tunable band gap (1.1-1.5 eV) which covers the optimal solar spectrum [38]. CTS compound is composed of three elements namely copper, tin and sulfur. These elements are nontoxic and earth abundant. CTS has been synthesized using both physical and chemical methods. Physical methods include co-sputtering, co-evaporation, PLD (pulsed laser deposition), sulfurization, ball mill while chemical methods include electrodeposition, SILAR (successive ionic layer adsorption and reaction), spin coating, spray pyrolysis, microwave irradiation, layer by layer solution method and sol gel method. The current trends in research is focused on the use of absorber compound for solar cell having reduced number of elements which are nontoxic and earth abundant and hence

CTS is suitable candidate for solar applications. The theoretical efficiency of CTS is 30%. KuKu and Fakolujo long back in 1987 first fabricated CTS solar cell with 0.11% PCE by direct evaporation method and hence following this, many reports based on CTS synthesis and its PCE were later reported. The current highest efficiency reported for CTS is 4.6% by Na doping and 6% by Ge doping [38].

Both physical and chemical techniques have their own advantages and disadvantages and it depends upon the application where they are adopted. Both techniques can be applied in solar cell fabrication. It's up to us to select any technique for absorber compound synthesis for solar application because in both the methods we cannot directly use the synthesized compound for solar cell fabrication as it requires post annealing treatments. The as-synthesized compound is either amorphous, non-compact or composition deficient and hence these factors affect the PCE of solar cells. The post annealing treatment is very important and it helps to overcome these problems. Based on the requirement of the desired properties in the absorber compound, annealing is done in vacuum, sulfur (sulfurization) or in selenium (selenization) environment.

5.4.2. Optical, structural and electrical properties of CTS

Optical study is very important as it determines the suitability of a material for photovoltaic (PV) application. In solar cell application, absorbance and band gap are the two important factors that decide the future application of material [39]. The band gap of CTS lies in the tunable range of 1.1-1.5 eV and has large absorption coefficient of 10^5 cm^{-1} [39]. The tunable property of band gap is due to quantum size effect arising due to confinement of electron and hole in a smaller volume. Band gap of a material can also be tuned by alloying. Alloying refers to doping with impurity which generates additional energy level leading to generation and separation of charge carriers.

CTS compound is considered to be bimorphic or polymorphic [40]. The phase formation depends upon the arrangement of cations and temperature. The tetragonal and monoclinic are the low temperature (750 ℃) stable structures while the cubic is the high temperature (750 ℃) metastable structure. The high temperature CTS phase is isomorphic with the cubic ZnS structure while the low temperature CTS phases are isomorphic with the tetragonal stannite $Cu_2(Fe,Zn)SnS_4$ structure [41]. In cubic CTS structure, 2/3 and 1/3 Zn^{2+} cations in ZnS structure are replaced by Cu^+ and Sn^{4+} cations and these cations have the same position as they have in ZnS structure [38]. In the tetragonal CTS structure and the composite atoms M1 [43.6(2) at % Sn + 56.4 at % Cu] and M_2 [46.3(3) at % Sn + 53.7(3) at % Cu] occupy the Cu and Sn atoms positions of stannite and the remaining Cu atoms in CTS occupy corresponding (Fe, Zn) atoms. The crystal structure of monoclinic CTS structure has close cubic packing of S atoms with Cu and Sn in tetrahedral sites. There is very slight difference between cubic and monoclinic CTS lattice parameters values. Hence, CTS tetragonal is the intermediate phase between high temperature cubic and low temperature monoclinic phase. These ternary compounds exhibit semiconducting property due to lack of direct S-S bonding, [42]. The reported band gap values for cubic, tetragonal and monoclinic are 1.0, 1.28 and 0.92 eV, respectively [43, 44]. The different band gaps of these structures are due to different geometrical arrangements of the cations and anions in the atomic structures and these band gap values can be tuned by doping (impurity addition).

The structure of CTS is also affected by the Cu/Sn compositional ratio. In stoichiometric CTS compound, the number of cations and anions is same, while in case of Cu rich compound, the number of cations exceeds the number of anions. Fig. 5.14 indicates the calculated chemical potential stability of Cu-Sn-S compound where it is seen that Cu_2SnS_3 has the widest potential range [45]. As seen from the figure, Cu_2SnS_3 at Cu rich site is connected to Cu_2S phase through Cu_4SnS_4 phase and to CuS phase through Cu_3SnS_4 phase. Thus, it is understood that when the composition is Cu rich, there is a maximum probability of forming secondary phases of CuS, Cu_2S, Cu_3SnS_4 and Cu_4SnS_4. On the other hand when the composition is Cu poor, SnS, SnS_2 and $Cu_4Sn_7S_{16}$ phases are formed. Hence, the CTS structure formation is also influenced by the elemental composition.

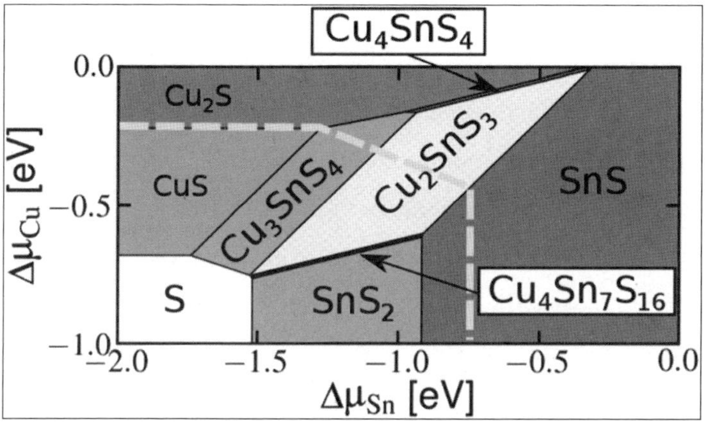

Fig. 5.14. Calculated chemical potential phase space of Cu-Sn-S system. Reprinted with permission from reference [45].

The secondary phases in CTS such as CuS and SnS significantly affect the optical properties of CTS compound. These phases act as defects in the crystal structure and hence they affect the optical properties such as band gap and transmittance of the material. The band gap is changed from 2.19 to 1.77 eV due to the presence of SnS secondary phase at higher annealing temperature (425℃). The secondary phases in CTS results in low transmittance. Similarly, presence of CuS secondary phase leads to low transmittance.

The electrical properties of CTS thin films are also influenced by the secondary phases. Copper rich phases such as Cu_3SnS_4, Cu_4SnS_4 and Cu_2S are conductive and hence their existence in the compound can lead to increased conductivity [45]. Even though Cu_4SnS_4, Cu_2SnS_3 and $Cu_4Sn_7S_{16}$ have same electronic structure, only Cu_2SnS_3 is suitable for solar cell application because $Cu_4Sn_7S_{16}$ has maximum probability of charge trapping due to narrow band high density of states and Cu_4SnS_4 has high electrical conductivity which is not suitable for photovoltaic devices [46]. KCN etching can be employed for the removal of conductive CuS phase from CTS compound. Hence, secondary phases play a vital role in modifying the optical and electrical properties of CTS thin films and so the PCE of solar cells. The oxidation state of Cu in CTS compound also plays a very important role in controlling the electronic property. The presence of Cu in +2 state is metallic state and will cause shunt losses in the solar cell therefore Cu must be present in +1 oxidation state.

CTS absorber based thin film solar cells have shown reliable photovoltaic efficiency. Table 1 shows the reported PCE of CTS based solar cells. From the reported solar efficiency, highest efficiency is reported for vacuum evaporation technique by Na doping. The efficiency was increased from 0.43 to 4.63 % with V_{oc} 283 mV with increased Na doping (0.75 NaF/Cu). This effect of increasing V_{oc} is attributed to the increased carrier concentration due to Na doping making the thermal activation energy lower and pulling the fermi level towards the valance band which leads to higher built in potential. The reported highest efficiency for CTS solar cells 6.01 % with V_{oc} 0.355 V, J_{sc} 30.5 mA/cm^2 and FF 0.556 by Ge doping [47]. The enhanced V_{oc} and J_{sc} is due to increased carrier life time, reduced carrier concentration and enhanced crystallinity of CTS films by Ge doping which favored enhanced efficiency of the solar cell.

Table. 1. The reported highest PCE of CTS thin film solar cells by various techniques.

Reference	Method	Efficiency (%)
48	Dip coating	2.10
48	Electrodeposition	2.84
48	Chemical bath deposition	0.961
49	Vacuum Evaporation	4.63
44	Sputtering	3.05
50	E Beam Evaporation	2.7
51	Ball Milling	1.92

CTS is secondary phase during formation of CZTS compound. Hence, research was focused to explore the properties of CTS compound and now solar cells with enhanced efficiencies (4.63% and 6.01%) have been reported. During CTS formation, CuS and SnS are formed as secondary phases. These binary semiconductor compounds also exhibit the desired properties suitable for solar cell application. Hence, these materials have been well researched and good efficiency upto 4.6% for SnS [52] and 10% for Cu_2S [53] have been reported. Thus, these binary compounds can be reliably applied for photovoltaic applications

5.4.3. CTS solar cell device fabrication

The schematic of CTS solar cell is shown in Fig. 5.15. CTS

Thin film solar cells are fabricated in tandem structure. It consists of molybdenum (Mo) layer (1 μm) sputter coated on soda lime glass substrate (SLG). The absorber material (p type-CTS) is then coated over Mo/SLG substrate within the thickness range of 1-1.2 μm by either physical or chemical techniques. The composition, crystal structure, and the microstructural properties of the absorber layer are modified at this step by sulfurization. The buffer layer ('n' type) is then deposited by chemical bath deposition (CBD) method. Usually, CdS (bandgap 2.4 eV) is used as a buffer layer. The CdS thickness is low (40-70nm) in order to keep the low series resistance. After buffer layer, intrinsic i-ZnO (50-70 nm) is sputtered followed by final sputtering of ZnO transparent conducting oxide (TCO) layer with metallic Al/Ag electrical contact grids. Fig. 5.15 shows the schematics of CTS solar cells.

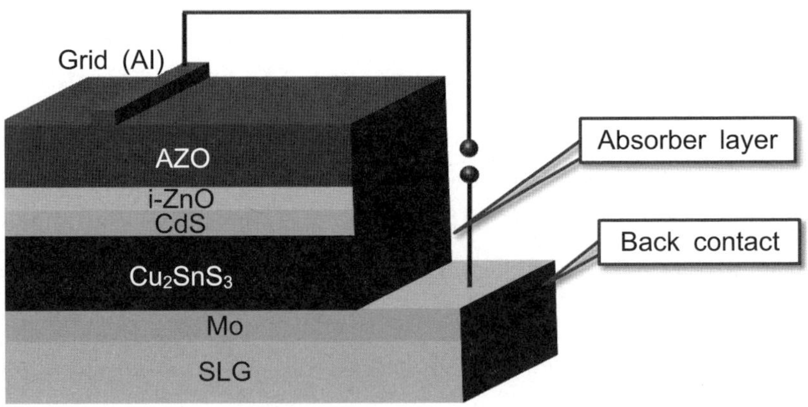

Fig. 5.15. Schematic of CTS solar cell.

The important step in CTS solar cell device fabrication is the sulfurization step. In this step, the absorber compound (CTS) is annealed at certain temperature (400-500℃) for a certain time period (5-20 min) to induce the desired properties in absorber required for solar cell. The absorber compound needs to be compact and highly crystalline to allow easy passage of electrons and avoid recombination losses at the non-compact interfaces. The absorber composition also needs to be in desired stoichiometry for solar cell application and hence is achieved by this step. Sulfurization of CTS film is conducted in a furnace in an inert (N_2 or Ar) or air atmosphere. Usually, elemental sulfur powder (5-50 mg) or H_2S gas is used for sulfurization. Sulfurization is also done in rapid thermal annealing (RTA) furnace where the absorber compound is heated very rapidly (ramping, 10-100℃/min) to the desired temperature, hold for some time and then cooled to room temperature.

References

1. Antonio Luque and Steven Hegedus John Wiley & Sons Ltd, "Handbook of photovoltaic science and engineering", 2003 (Book).
2. Zhang S, Wei S, Zunger A, Katayama-Yoshida H, Phys. Rev. B 57, 9642-9656 (1998).
3. Yoshihiro Hamakawa, Thin-film Solar cells Next Generation Photovoltaics and its Applications. Spring series in Photonics, 2003.
4. Martin A. Green, Keith Emery, Yoshihiro Hishikawa, Wilhelm Warta and Ewan D. Dunlop, "Solar cell efficiency tables (version 47), Prog. Photovolt: Res. Appl. 24, 3-11 (2016).
5. Junker, S.T., "Modeling and control of a continuous co-evaporative physical vapor deposition process for production of thin-film copper indium gallium diselenide photovoltaic cells". Ph.D thesis, Univeristy of Delaware (2003).
6. Solar Frontier Achieves World Record Thin-Film Solar Cell Efficiency 22.3%.
7. Fahrenbruch A, Bube R, Kim D, Lopez-Otero A, Int. J. Sol. Energy 12, 197-222 (1992).
8. First Solar raises bar for CdTe with 21.5% efficiency record, 6 February 2015, PV magazine.
9. K. Ito and T. Nakazawa, "Electrical and optical properties of stannite-type quaternary semiconductor thin films", Jpn. J. Appl. Phys. 27, 2094 (1988).

10. H. Katagiri, N. Sasaguchi, S. Hando, S. Hoshino, J. Ohashi, T. Yokota, "Preparation and evaluation of Cu_2ZnSnS_4 thin films by sulfurization of E - B evaporated precursors", Sol. Energy Mater. Sol. Cells, 49, 407 (1997).

11. B. R. Pamplin, "Super-Cell Structure of Semiconductors", Nature, 188, 136 (1960).

12. S. Chen, X. G. Gong, A. Walsh, S.-H. Wei, "Electronic structure and stability of quaternary chalcogenide semiconductors derived from cation cross-substitution of II-VI and I-III-VI2 compounds", Phys. Rev. B, 79, 165211 (2009).

13. S. Schorr, "Structural aspects of adamantine like multinary chalcogenides", Thin Solid Films, 515, 5985 (2007).

14. U. Ghorpade, M. Suryawanshi, S. W. Shin, K. Gurav, P. Patil, S. Pawar, C. W. Hong, J. H. Kim, S. S. Kolekar, "Towards environmentally benign approaches for the synthesis of CZTSSe nanocrystals by a hot injection method: a status review", Chem. Commun. 50, 11258 (2014).

15. S. Chen, X. G. Gong, A. Walsh, S.-H. Wei, "Crystal and electronic band structure of Cu_2ZnSnX_4 (X = S and Se) photovoltaic absorbers: First-principles insights", Appl. Phys. Lett. 94, 041903 (2009).

16. M. P. Suryawanshi, G. L. Agawane, S. M. Bhosale, S. W. Shin, P. S. Patil, J. H. Kim, A. V. Moholkar, "CZTS based thin film solar cells: a status review", Mater. Tech. 28, 98 (2013).

17. S. A. Vanalakar, G. L. Agawane, S. W. Shin, M. P. Suryawanshi, K. V. Gurav, K. S. Jeon, P. S. Patil, C. W. Jeong, I. Y. Kim,

J. H. Kim, "A review on pulsed laser deposited CZTS thin films for solar cell applications", J. Alloys Compd. 619, 109 (2015).

18. M. Leon, S. Levcenko, R. Serna, A. Nateprov, G. Gurieva, J. M. Merino, S. Schorr, E. Arushanov, "Spectroscopic ellipsometry study of $Cu_2ZnGeSe_4$ and $Cu_2ZnSiSe_4$ poly-crystals", Mat. Chem. Phys. 141, 58 (2013).
19. Y. Yang, X. Kang, L. Huang, D. Pan, "Tuning the band gap of $Cu_2ZnSn(S, Se)_4$ thin film via lithium alloying", ACS Appl. Mater. Interfaces 8, 5308 (2016).
20. S. Chen, X. G. Gong, A. Walsh, S. H. Wei, "Defect physics of the kesterite thin-film solar cell absorber Cu_2ZnSnS_4", Appl. Phys. Lett. 96, 021902 (2010).
21. S. Chen, A. Walsh, X. G. Gong, S. H. Wei, "Classification of lattice defects in the kesterite Cu_2ZnSnS_4 and $Cu_2ZnSnSe_4$ earth-abundant solar cell absorbers", Adv. Mater. 25, 1522 (2013).
22. S. Chen, J. H. Yang, X. G. Gong, A. Walsh, S. H. Wei, "Intrinsic point defects and complexes in the quaternary kesterite semiconductor Cu_2ZnSnS_4", Phys. Rev. B 81, 245204 (2010).
23. R. A. Wibowo, E. Soo Lee, B. Munir, K. Ho Kim, "Pulsed laser deposition of quaternary $Cu_2ZnSnSe_4$ thin films", Phys. Status Solidi (B) 204, 3373 (2007).
24. T. Tanaka, T. Nagatomo, D. Kawasaki, M. Nishio, Q. Guo, A. Wakahara, A. Yoshida, H. Ogawa, "Preparation of Cu_2ZnSnS_4 thin films by hybrid sputtering", J. Phys. Chem. Solids 66, 1978 (2005).
25. J. J. Scragg, P. J. Dale, L. M. Peter, "Towards sustainable materials

for solar energy conversion: Preparation and photoelectrochemical characterization of Cu_2ZnSnS_4", Electrochem. Commun. 10, 639 (2008).

26. V. Chawla and B. Clemens, "Effect of Composition on High Efficiency CZTSSe Devices Fabricated Using Co-sputtering of Compound Targets", Proc. of the 38th IEEE Photovoltaic Specialists Conference, Austin, Texas, USA (2012).

27. J. Kim, H. Hiroi, T. K. Todorov, O. Gunawan, M. Kuwahara, T. Gokmen, D. Nair, M. Hopstaken, B. Shin, Y. S. Lee, W. Wang, H. Sugimoto, D. B. Mitzi, "High efficiency $Cu_2ZnSn(S,Se)_4$ solar cells by applying a double In_2S_3/CdS emitter", Adv. Mater., 26, 7427 (2014).

28. H. Hiroi, N. Sakai, T. Kato and H. Sugimoto, "High voltage Cu_2ZnSnS_4 submodules by hybrid buffer layer", in Photovoltaic Specialists Conference (PVSC), 2013 IEEE 39th, 0863-0866 (2013).

29. J. Li, H. Wang, M. Luo, J. Tang, C. Chen, W. Liu, F. Liu, Y. Sun, J. Han, Y. Zhang, "10% Efficiency $Cu_2ZnSn(S,Se)_4$ thin film solar cells fabricated by magnetron sputtering with enlarged depletion region width", Sol. Energy Mater. Sol. Cells, 149, 242 (2016).

30. S. W. Seo, J. O. Jeon, J. W. Seo, Y. Y. Yu, J. H. Jeong, D. K. Lee, H. G. Kim, M. J. Ko, H. J. Son, H. W. Jang, J. Y. Kim, "Compositional and interfacial modification of $Cu_2ZnSn(S,Se)_4$ thin-film solar cells prepared by electrochemical deposition", ChemSusChem. 9, 439 (2016).

31. I. Repins, C. Beall, N. Vora, C. DeHart, D. Kuciauskas, P. Dippo, B. To, J. Mann, W.-C. Hsu, A. Goodrich, R. Noufi, "Co-evaporated $Cu_2ZnSnSe_4$ films and devices", Sol. Energy Mater. Sol. Cells 101, 154 (2012).

32. C. K. Miskin, W. C. Yang, C. J. hages, N. J. Carter, C. S. Joglekar, E. A, Stach, R. Agrawal, "9.0% efficient $Cu_2ZnSn(S, Se)_4$ solar cells from selenized nanoparticle inks", Prog. Photovolt: Res. Appl. 23, 654 (2015).

33. I. Olekseyuk, I. Dudchak, L. Piskach, "Phase equilibria in the Cu_2S - ZnS -SnS_2 system", J. Alloys Compd. 368, 135 (2004).

34. Carolin Maria Sputter-Fella, "Solution-processed kesterite absorbers or thin film solar cells", Friedrich-Alexander-Universitat Erlangen-Nurnberg, Germany, Ph. D. Thesis, 2014.

35. J. J. Scragg, P. J. Dale, D. Colombara, L. M. Peter, "Thermodynamic aspects of the synthesis of thin-film materials for solar cells", ChemPhys Chem. 13, 3035 (2012).

36. J. J. Scragg, J. T. Watjen, M. Edoff, T. Ericson, T. Kubart and C. Platzer-Bjorkman, "A detrimental reaction at the molybdenum back contact in $Cu_2ZnSn(S,Se)_4$ thin-film solar cells", J. Am. Chem. Soc. 134, 19330 (2012).

37. X. Liu, Y. Feng, H. Cui, F. Liu, X. Hao, G. Conibeer, D. B. Mitzi, M. A. Green, "The current status and future prospects of kesterite solar cells: a brief review", Prog. Photovolt: Res. Appl. 24, 879 (2016).

38. A.C. Lokhande, K.V. Gurav, Eunjin Jo, C.D. Lokhande, Jin Hyeok Kim, J. Alloys Compd. 656, 295-310 (2016).

39. P. Zhao and S. Cheng, Adv. Mater. Sci. Eng., 2013, 1-4 (2013).

40. M. Onoda, X. Chen, A. Sato and H. Wada, Mater. Res. Bull, 35, 1563-1570 (2000).

41. S.R. Hall, J.T. Szymanski and J.M. Stewart, Can Mineral, 16, 13-137(1978).

42. X. Chen, H. Wada, A. Sato, and M. Mieno, J. Solid State Chem., 139, 14 -151(1998).

43. R. Chierchia, F. Pigna, M. Valentini, C. Malerba, E. Salza, P. Mangiapane, T. Polichetti,and A. Mittig, Phys. Status Solidi C, 13, 35-39(2016).

44. Y. Dong, J. He, X. Li, W. Zhou,Y. Chen, L. Sun, P. Yang, J. Chu, Mater. Lett., 160, 468-471(2015).

45. P. Zawadzki, L. L. Baranowski, H. Peng, E. S. Toberer, D. S. Ginley, W. Tumas, A. Zakutayev, and S. Lany, Appl. Phys. Lett., 103, 253902 (2013).

46. L. L. Baranowski, P. Zawadzki, S. Christensen, D. Nordlund, S. Lany, A. C. Tamboli, L. Gedvilas, D. S. Ginley, W. Tumas, E. S. Toberer and A. Zakutayev, Chem. Mater., 26, 4951-4959 (2014).

47. M. Umehara, Y. Takeda, T. Motohiro, T. Sakai, H. Awano, and R. Maekawa, APPL PHYS EXPRESS, 6, 045501 (2013).

48. J. Li, C. Xue, Y. Wang, G. Jiang, W. Liu, C. Zhu, Sol. Energy Mater. Sol. Cells, 144, 281-288 (2016).

49. M. Nakashima, J. Fujimoto, T. Yamaguchi, and M. Izaki, Appl Phy Express, 8, 042303 (2015).

50. N. Aihara, A. Kanai, K. Kimura, M. Yamada, K. Toyonaga, H.

Araki, A.Takeuchi, and H. Katagiri, Jpn. J. Appl. Phys., 53, 05FW13 (2014).

51. Q. Chen, X. Dou, Y Ni, S. Cheng, S. Zhuang, J. Colloid Interface Sci., 376 (2012) 327-330.
52. V. R. M. Reddy, S. Gedi, C. Park, Miles R.W, Ramakrishna Reddy K.T, Curr. Appl Phys., 15, 588-598 (2015).
53. Scudder, L. R.; Godlewski, M. P.; Bozek, J. M.; Klucher, T. M.Shorting, NASA TN D-6362; National Aeronautics and Space Administration: 1971.

Chapter 6
Analysis and characterization methods for solar cells

6.1. Introduction

In the previous chapters we have learnt the basics of semiconductors and fabrication of thin film solar cells. The performance of the solar cells strongly depends on various factors. Understanding the correlation between process conditions on the parameters of the solar cells is essential to continuously improve the efficiency of solar cells and further to identity new materials for solar cell applications. One of the important factors of development of solar cells is achieving high conversion efficiency and low cost production methods. And there strong correlation between the material properties such as life time, diffusion length, grain boundaries, defects and impurities on the photovoltaic performance of the solar cells. Therefore, analysis of solar cell parameters is prerequisite for further improvement and development highly efficient solar cells. This chapter briefly outlines the measurement techniques used for characterization and analysis of solar cells.

6.2. Illuminated current-voltage characteristics

The most important measurement of completed solar cell is illuminated current - voltage characteristics. The measurement was performed using a solar simulator. The schematic diagram, and system image of the measurement system is shown in the figures 6.1 and 6.2, respectively. The measurement system consists of Xenon lamp as light source. It is known that the Xenon lamp gives very broad white spectrum; this is corrected using filters to obtained desired spectrum equivalent to the AM 1.5. Air mass spectrum is an international standard for terrestrial cells and its intensity is 100 mW/cm^2 (1 kW/m^2, one-sun). The measurements were carried out at room temperature of 25°C. A four probe configuration is generally used to reduce the losses of the contact resistances of the probes.

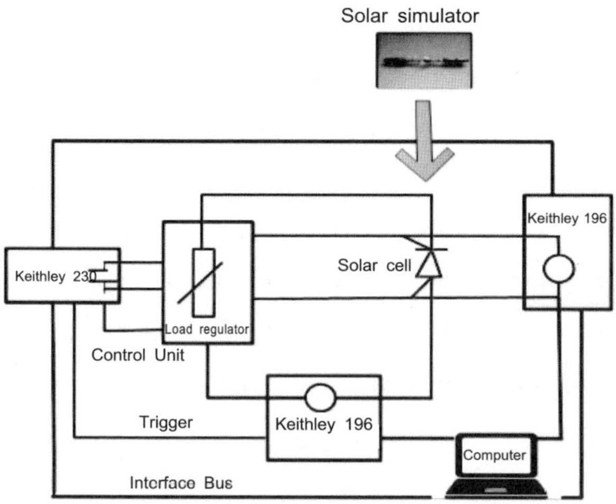

Fig. 6.1. Schematic block diagram of solar simulator.

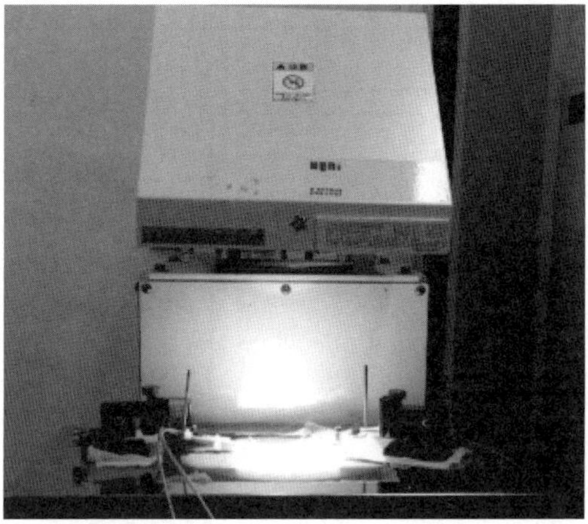

Fig. 6.2. Photo of solar simulator for I-V measurement.

The I-V curve of the solar cell is shown in Fig. 6.3. The curve gives solar cell parameters such as open circuit voltage, short circuit current J_{sc}, and fill factor FF and conversion efficiency. In addition to this power maximum values for current and voltages are also extracted from the figure. The four parameters can be used to evaluate the solar cell performance.

As mentioned in pervious chapters that solar cell parameters were depend on semiconductor junction properties. Open circuit voltage V_{oc} depends on the saturation current or dark current, minority carrier diffusion length L_n and minority carrier life times. The fill factor, measure of square of the I-V curve depends on parasitic resistances R_s and R_{sh}.

While short circuit density J_{sc} is depends on reflection losses due to shadowing of electrical contacts and incomplete absorption of long wavelength radiation must be understood to reduces the

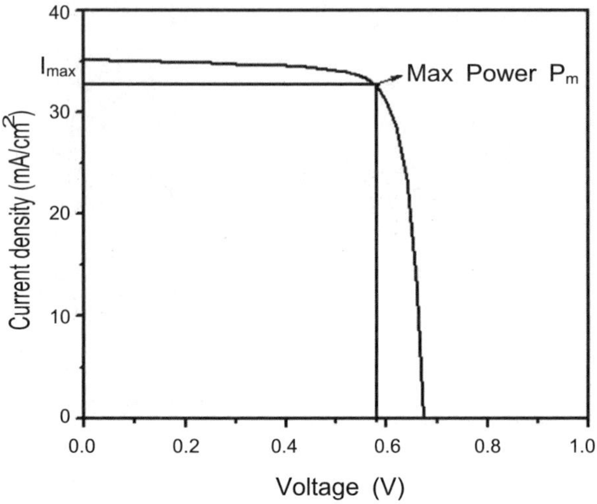

Fig. 6.3. The illuminated current-voltage characteristics of CIGS solar cell.

J_{sc} losses.

In thin film solar cells open circuit voltage is reduced due to recombination of carriers. As we learnt from the previous chapters that recombination occurs in three ways (1) radiative recombination, (2) non radiative recombination and (3) Auger recombination. In general, all thin film solar cells CIGS, CdTe and CZTSSe exhibit non radiative recombination via traps which is known as Shockley-Read-Hall recombination [1].

By measuring the V_{oc} as function of temperature one can understand the dominant recombination mechanism in the solar cells. Fig. 6.4 shows typical curve of open circuit voltage versus temperature of CIGS thin film solar cell. By extrapolating the V_{oc} curve one can estimate the activation energy E_a. If the E_a is equal to the band gap then the recombination is due to space charge recombination via traps, On the other hand if activation energy E_a is lower than the band gap of the absorber indicates recombination at the interface between absorber and buffer layers.

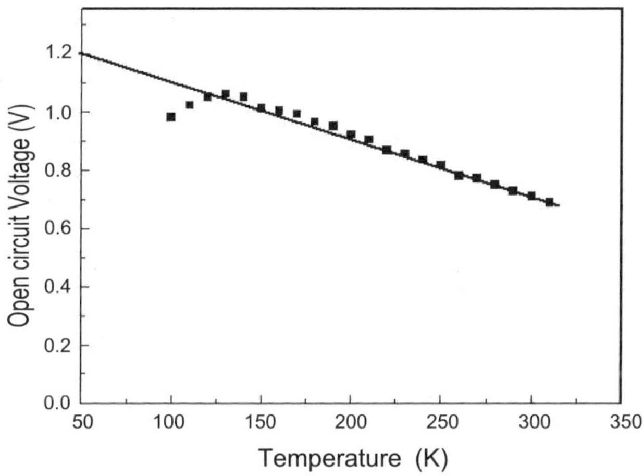

Fig. 6.4. Open circuit voltage versus temperature (K) of a CIGS solar cell.

$J_{sc}V_{oc}$ and V_{oc}-Suns measurement: In another way of measuring the I-V are J_{sc}-V_{oc} and suns-V_{oc} method. In this method the illumination intensity is varied and corresponding J_{sc} and V_{oc} are measured. This measurement can be performed all kinds solar cells. The measurement system consists of flash lamp and filters, removing the filters the intensity can be varied from 1 sun to 300 suns. Typically two probes used to measure the V_{oc}. The light intensity and V_{oc} are measured. From the V_{oc} versus intensity curves is easier to fit since there is no series resistance compared to dark and illuminated I-V. The ideality factor 'n' and be evaluated from the Suns V_{oc} curves. [2].

The ideality factor is estimated from the following relation

$$n_s = \left[\frac{V_T \, dInJ_{SC}}{dV_{oc}} \right]^{-1} \tag{6.1}$$

6.3. Dark current-voltage characteristics

From the light I-V characteristics it is difficult to understand the actual losses in solar cell due to variation of parasitic resistances with illumination. In addition to illuminated characteristics dark I-V measurements were performed on solar cell devices to understand the actual behavior of the junction. A typical dark I-V curve is given in Fig. 6.5 and log J_o versus voltage is shown in Fig. 6.6.

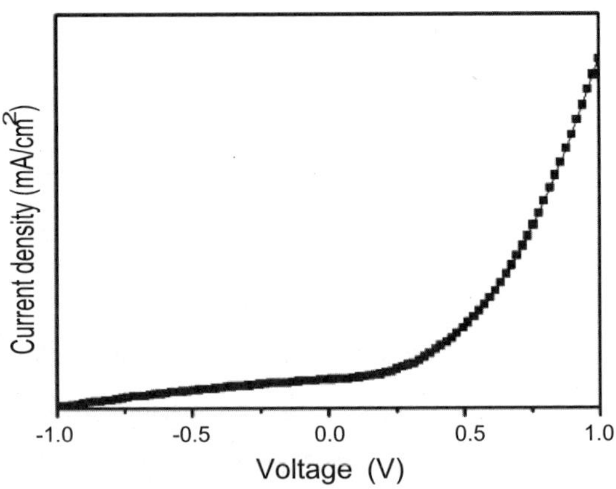

Fig. 6.5. The dark current-voltage characteristics of a solar cell.

The curve is fitted to the double diode equation given in equation [3],

$$I(V) = I_{01}\left[exp\frac{V-IR_s}{n_1 V_T}-1\right]+I_{02}\left[exp\frac{V-IR_s}{n_2 V_T}\right]+\frac{V-IR_s}{R_p} \quad (6.2)$$

At low voltage region of the figure shows dominance of the shunt conductance R_{sh} and at higher voltages series resistance R_s dominates.

The equation has two ideality factors n_1 and n_2, the diode ideality factor n_1 originates due to higher voltage region and n_2 is from lower voltage dependence. The most solar cells have dependency on n_1, however in some cases when the block contacts observed in solar cells it can be modeled as second diode and n_2 has significant influence on the performance of the solar cells.

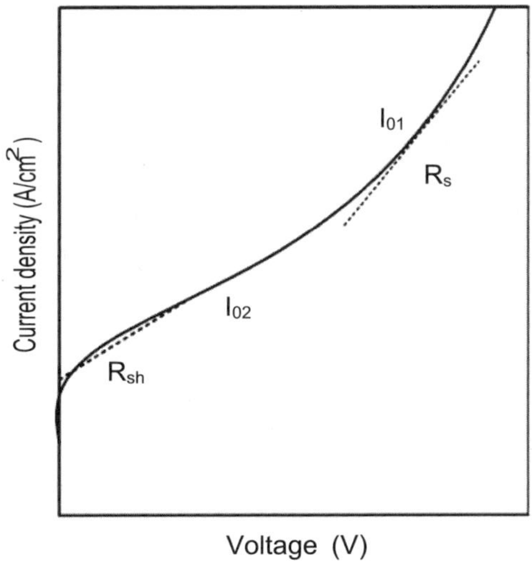

Fig. 6.6. log current-voltage characteristics of a solar cell.

The ideality factors n_l values of most solar cells falls in the range 1 to 2 indicating the recombination at the space charge region of the solar cells. Hence, measuring dark I-V currents will shed important physical mechanism limiting the performance of the solar cells. The typical values for n_l for silicon, thin film solar cells CIGSe, CdTe and CZTSSe are between 1 and 2.

The series resistance and parallel reshistance can be evaluated from the following relations

$$V_a - V_{oc} = R_s I_{sc} \tag{6.3}$$

and

$$R_S = \frac{V_a - V_{oc}}{I_{sc}} \tag{6.4}$$

In thin film solar cells such as CIGSe, CdTe and CZTSSe, R_s and shunt conductance G_{sh}, diode ideality factor, saturation dark current can evaluated from the curves of illuminated and dark I-V curves using procedure developed to evaluate the losses in thin film solar cells [4]. Series resistance arises from intrinsic resistances of the back contact, base or absorber, emitter or buffer layer and top contacts. High R_s reduce the J_{sc} there by reducing the achievable conversion efficiencies.

The solar cell shows lower performance with reduced shunt resistance. The shunt in solar cells originates for non-uniformities, leaky p-n junction. Shunt resistance should be as high as possible to reduce losses due V_{oc}. It is impossible to eliminate these resistances. However, they are minimized to lowest level to reduce its impact on fill factor FF and on solar cell efficiency.

The equation for single diode can be written as

$$I = I_o\left(exp\frac{qV}{AKT} - 1\right) \tag{6.5}$$

Where I_0 is dark current or saturation current, it is related to semiconductor parameters

$$I_0 = \left(\frac{qD_n n_i^2}{L_n N_A} + \frac{qD_p n_i^2}{L_p N_D}\right) \tag{6.6}$$

Where q is the electric charge, D_n and D_p are diffusion coefficient for electrons and holes, respectively. N_A and N_D are carrier concentration of acceptor and donors, respectively. The n_i^2 is the intrinsic carrier concentration; L_n and L_p are diffusion lengths for electrons and holes, respectively. It is clear from the above equation that electron diffusion length and minority carrier life time should be as high as possible to achieve low saturation current density. Lower saturation current densities of solar cells exhibit higher conversion efficiencies. So it is important to measure life time and diffusion lengths in order to reduce the dark current in solar cells.

6.4. Quantum efficiency measurements

Another important characterization technique for solar cells is measurement of the spectral response as a function of wave length of solar cells. The total current generated by solar cell is measured using Quantum efficiency measurements. The quantum efficiency measurement tells us how efficiently the solar cells converts incident light of a specific wavelength into photocurrent. The obtained value is the ratio of the number of charge carriers collected to the number of photons in the incident light. Fig. 6.7 shows schematic diagram of quantum efficiency measurement used for determine the spectral response of solar cells.

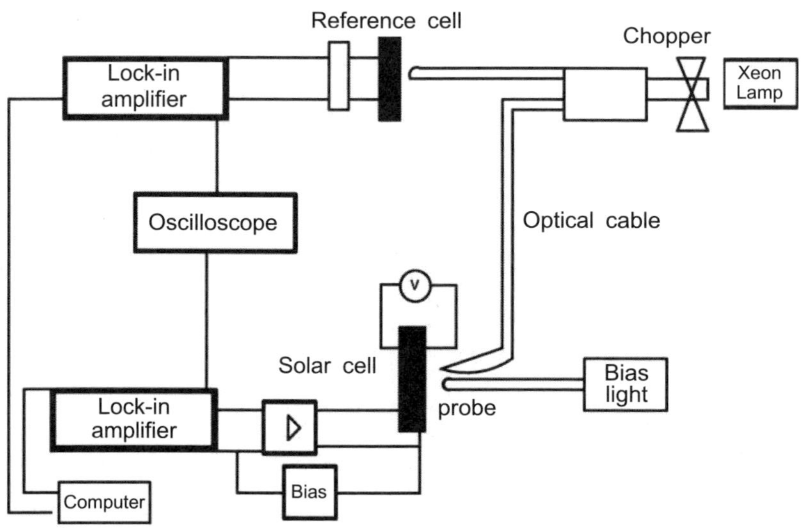

Fig. 6.7. Schematic block diagram of Quantum efficiency measurement.

The illumination system consists of a Xenon lamp; the light from the lamp is chopped, filtered and the beam is spiltted into two beams one is incident on reference solar cell and another half on the device under test. The short circuit current is measured with current sensitive lock-in amplifier. The EQE can be measured by applying light and voltage bias. In order measure the internal quantum efficiency of solar cell, the reflection must be included in the analysis.

There are two methods of quantum efficiency measurements were performed on solar cells, external quantum efficiency (EQE) and internal quantum efficiency (IQE). EQE gives the number of electrons for incident photons on the solar cell. On the other hand in IQE only photon that are absorbed by the solar cell, subtracting the reflectance of the cell is taken into consideration, Hence, IQE is always higher than EQE.

When exposed to illumination the solar cell produces current. The photo current is proportional to the incident radiation. The total current produced by the solar cells can be determined by integration of total incident solar AM 1.5 spectrum. Furthermore, from EQE measurements various losses which reduce J_{sc} in solar cells can be understood. Fig. 6.10 shows a typical External quantum efficiency EQE curve of a CIGS solar cell.

In the above figure it is also shown various losses due to window layer, buffer layer and reflection from the front surfaces of a CIGS solar cell. It can be seen at low wavelength region the absorption is from ZnO (3.4 eV) and absorption at CdS (2.41 eV) reduces available photocurrent.

As shown in Fig. 6.8, the losses due to shadowing and reflection from the top contacts showed at point 1 and 2, respectively. In the long wavelength region incomplete generation of carriers 5 and collection of carriers at point 6 reduces the total

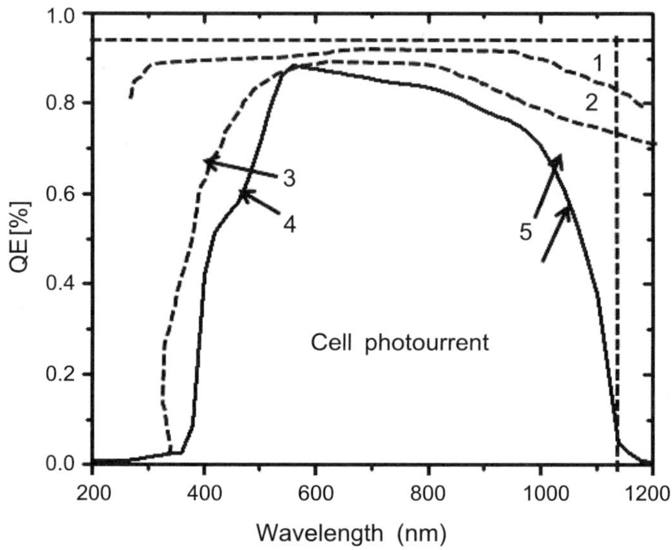

Fig. 6.8. EQE curve of a typical CIGS solar cell with losses.

available photocurrent from the incident spectrum. The total available photocurrent of typical CIGS solar cells is given by the following equation

$$I_{L_{total}} = \int_{\lambda min}^{\lambda max} I_L(\lambda) d\lambda \tag{6.7}$$

Fig. 6.9 shows the losses in CdTe thin film solar cells. As can be seen from the figure that various losses can be estimated due to various layers in solar cells which can be used to adjust the parameters of the layers to achieve higher conversion efficiencies of thin film solar cells.

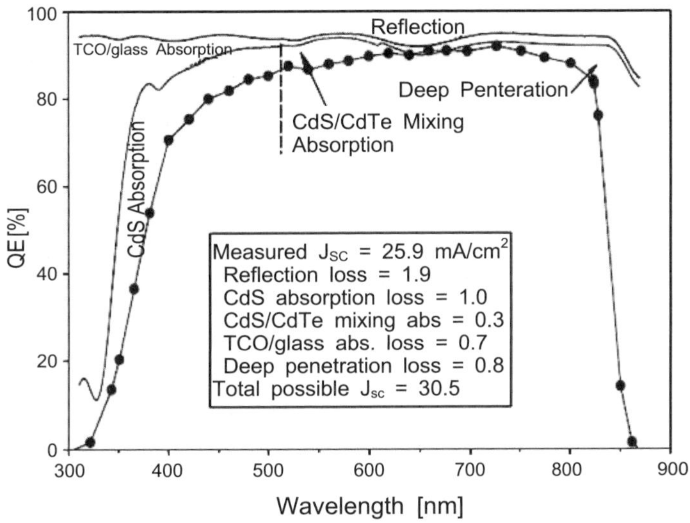

Fig. 6.9. EQE curve of a typical CIGS solar cell with losses [5].

QE measurement can also be used to measure spectral response of solar cells. Spectral response is similar to quantum efficiency. Spectral response is defined as ratio of photo generated current to incident solar power. The spectral response can be estimated from the QE from the following relation.

$$SR = \frac{q\lambda}{hc} QE \tag{6.8}$$

It is clear that solar cell only absorbs photon energy equivalent to its band gap. The photons with energy above and lower than band gap are not absorbed. Photons with energy greater than band gap transfers energy to the carriers and thermalization occurs. Whereas lower energy photons passes through the cell without absorption.

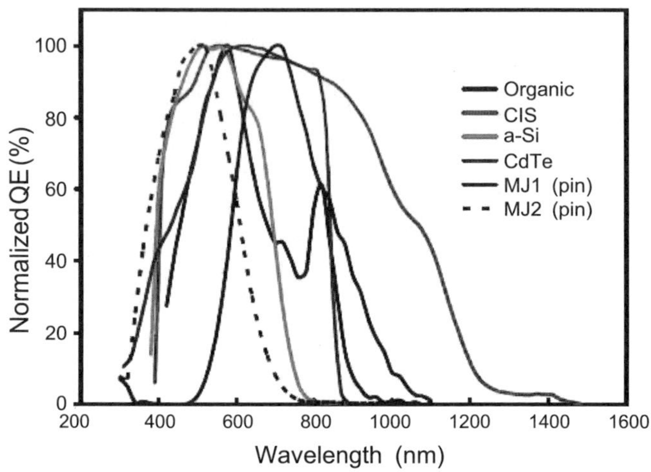

Fig. 6.10. Spectral response of different thin film solar cells [6].

Fig. 6.10 shows spectral response of three thin film materials. It can be seen from the spectral response of the CIGS thin film solar cell which has a broad response from 300 to 1,200 nm compared to other amorphous silicon and CdTe thin film and other solar cells. Its reveals that CIGS perform better in the diffuse and low light conditions compared to other solar cells. The spectral response is important measurement to understand the behavior of solar cells under various illumination conditions to shed light on the actual performance of the devices.

For a band gap of absorber about 1.2 eV the achievable current density is 46 A/cm^2. The single crystal silicon solar cells exhibits about 43 mA/cm^2 and maximum photo current for high efficiency CIGS solar cell is about 36 mA/cm^2 were achieved [7].

QE is also used to evaluate various buffer layers used for

CIGS thin film solar cells. Fig. 6.11 shows the QE curves of two CIGS solar cells with CdS and ZnS buffer layers. It can be seen from the figure that the QE curve of ZnS shows lower absorption losses at blue wavelength region of the device. This originates due to higher band gap of ZnS (E_g = 3.61 eV) to CdS (E_g = 2.41 eV) [4].

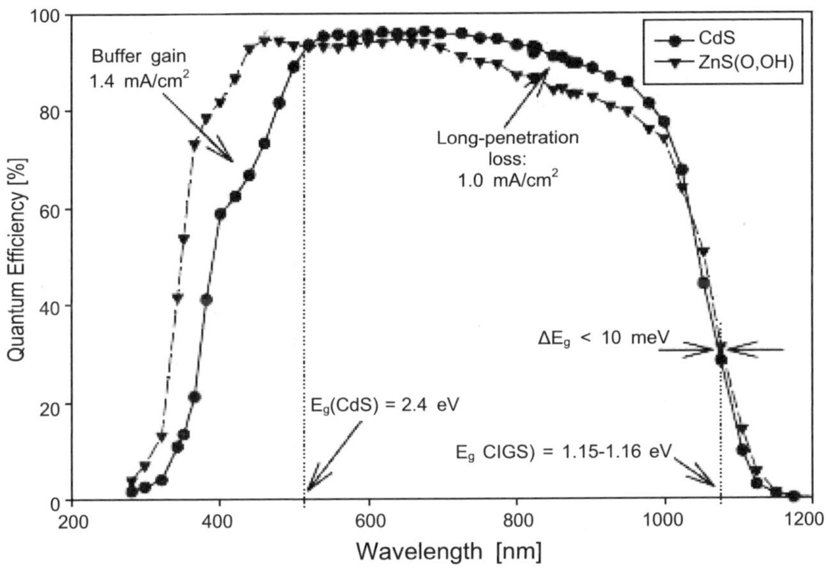

Fig. 6.11. EQE curve of a typical CIGS solar cell with losses [5].

However, at long wavelength region there is decrease in QE which suggests that the defects at interface between ZnS and CIGS thin film solar cells. In addition to this, measurement of EQE spectra is also used to determine the band gap the absorber. The plots of $[h\nu\text{-In }(1\text{-EQE})]^2$ versus $h\nu$ curves and long wavelength cut off gives rough estimate of the band gap of the absorber.

6.5. Measurement of diffusion length: EQE

The J_{sc} strongly depends on diffusion length of minority carriers; the J_{sc} is given following relation as function of L_n.

$$J_{sc} = qG(L_n + L_p + W) \tag{6.9}$$

Where q is electronic charge, G is the generation rate, the electron diffusion length L_n, hole diffusion lengths L_p and space charge region width W. So measurement of diffusion length of absorber is a useful for estimation of J_{sc}. The EQE measurements can used to estimate the diffusion length of the minority carriers. The internal quantum efficiency E_{int} is the ratio of photogenerated carriers collected to the incident photon flux on the absorber layer after reflection is given by the following expression.

$$QE_{int}(\lambda, V) = 1 - exp\frac{[-\alpha(\lambda)W(V)]}{\alpha L_n} + 1 \tag{6.10}$$

Where α is absorption coefficient of the absorber, W is the space charge width, and L_n is the minority carrier diffusion length. The above equation is valid for that for all carriers generated in the space charge region are collected without recombination loss. With known values of space charge width the diffusion length can be estimated utilizing the QE analysis. The above analysis can applied to all thin film solar cell based on CIGSe, CdTe and

CZTSSe materials. By measuring the quantum efficiency on solar cells one can extract various factors that are limiting available photocurrent, which in turn reduces the performance of the devices.

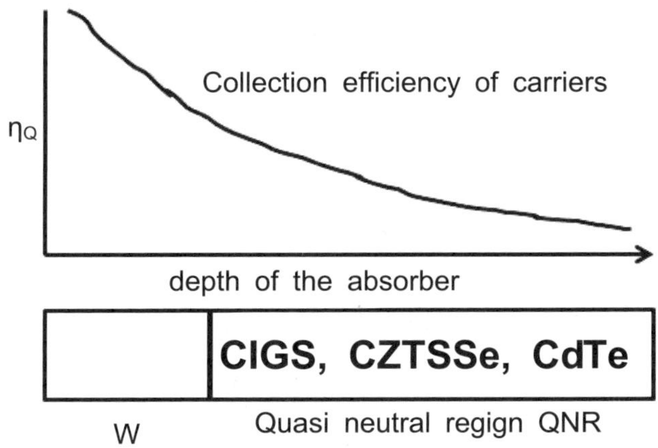

Fig. 6.12. EQE curve of a typical CIGS solar cell with losses.

Fig. 6.12 shows the collection efficiency of the absorbers as function of depth. It shows the collection efficiency decrease in the absorber. If the L is minority carrier diffusion length and 'w' width of the space charge region (0.6 μm) is $L_n + w = 0.5$-1.0 μm then all the carriers in this region is collected. In CIGS, CdTe and CZTSSe direct semiconductors it is about 1.0 μm, above this collection efficiency and J_{sc} decreases. Therefore all poly-crystalline semiconductors thin film solar cells have thickness about 1-4 μm which is sufficient for absorber thickness. Increasing beyond this the J_{sc} saturates and there is no improvement in the performance of the solar cells. As mentioned the L_n should be high for higher photocurrent. Hence, measurement of L_n is necessary to improve electronic quality of the absorbers used in thin film solar cells.

EQE can be also measured with voltage negative bias (-1V) and positive bias. Thin film solar cells exhibit the voltage depended current collection. The reduction in photogenerated current reduces the open-circuit voltage. From the above discussion reveals EQE is very important tool in analyzing the all kinds of solar cells to understand the current losses in the devices.

6.5.1. Diffusion length: EBIC

The minority carrier diffusion length is also estimated by electron beam induced current (EBIC). The EBIC can be performed on completed thin film solar cells of CIGS, CdTe and CZTSSe device. When electron beam is rastered across the device from top to bottom of the junction, the beam attenuates from junction due to presence of recombination centres originating from defects in the absorber.

The EBIC signal data is fitted after assuming that all the carriers generated in the depletion region of the device are collected and also the carriers generated outside of depletion region is also diffused into the depletion region. The beam current is given by following relation [8]

$$I(x) = I_0 \text{ for } x < w \tag{6.11}$$

And for x < w the beam signal is given by following relation

$$I(x) = I_0 exp\frac{(w-x)}{L_n} \text{For } x > w \tag{6.12}$$

Where I_0 is amplitude of measured signal, w is depletion width, L_n is minority carrier diffusion length and signal, x distance away from junction of absorber/buffer layer. Fig. 6.13 shows the EBIC signal fitted to the above equation.

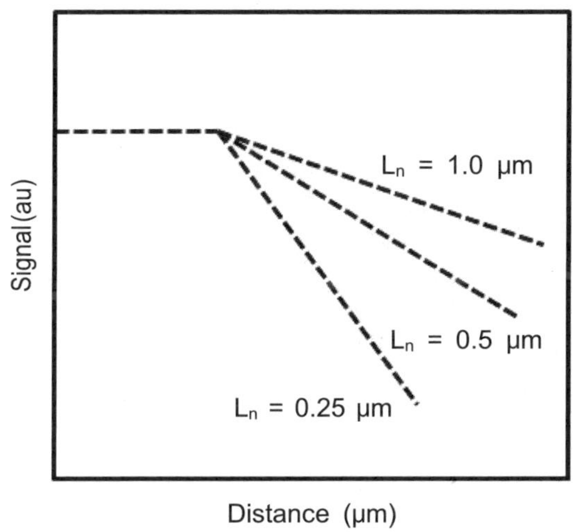

Fig. 6.13. Schematic EBIC signal fitted to the above equation.

6.6. Characterization of defects in solar cells

The presence of defects in the absorber strongly affects the performances of solar cell device performance. The defects in the structure can be analyzed using photoluminescence spectroscopy (PL). Deep level transient spectroscopy (DLTS), Admittance spectroscopy, and drive level capacitance profiling (DLCP) can be employed for determining the deep structural defects in the material. The minority carrier lifetime can be measured from time-resolved photoluminescence spectroscopy (TR-PL).

6.7. Photoluminescence (PL) spectroscopy

Photoluminescence (PL) spectroscopy is a contactless, nondestructive technique used to investigate defects, impurity levels of semiconductor materials. In this method a monochromatic light is incident onto the sample, after absorption of this light the photo exited carriers returns to ground state by emitting a characteristic spectrum.

The peak position reveals the band gap of the material. The intensity and spectral position of are used to identify the defect, impurity level energies in the forbidden band gap region. PL also used to understand the recombination mechanism in the chalcogenide semiconductors.

The typical photoluminescence spectra of CIGS, CZTSSe absorbers are shown in Fig. 6.14 Room temperature PL measurement is performed to identify the band gap of the CIGS, CZTS, CZTSe and CdTe thin films. To understand the recombination mechanism, temperature and excitation depended PL studies are generally performed. Generally, The PL emission in thin film materials are dominated by the donor-acceptor pairs (DAP), band to band (BB) or band to tail states (BT). At low doping levels the DAP is dominant and at higher doping levels

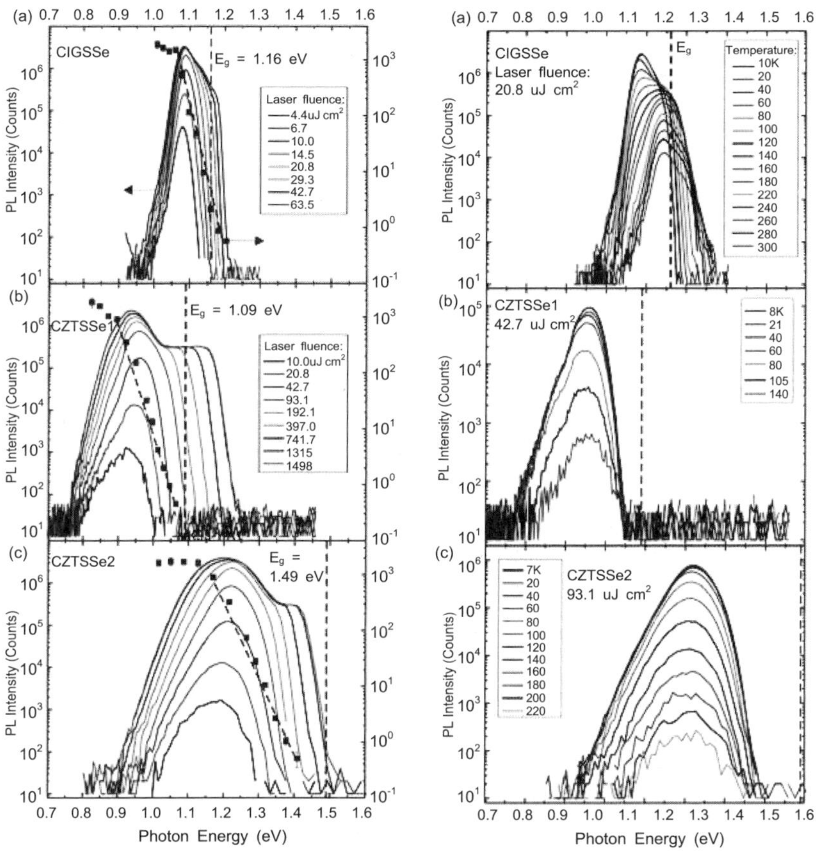

Fig. 6.14. Power and temperature depended Photoluminescence spectra of CIGS, CZTSSe absorbers [9].

due to potential fluctuation in the sample leads to band to tail states due impurities. The intrinsic shallow and deep defect energy levels are estimated in the CIGS and CZTSSe and other polycrystalline materials using Photoluminescence.

The PL peak intensity blue shifts with increase in excitation intensity. The intensity of the peak is given by the relation [9].

$$I = P^m \tag{6.13}$$

Where the I is excitation intensity, P is excitation power and m is variable, for recombination via defects the $m < 1$ and $m > 1$ involves recombination from band-to-band. With increase in excitation power, the defects are neutralized due to carrier trapping and free charge carriers increases which results in fluctuating potential which results in blue shift in the spectrum. With increase in temperature the PL intensity is quenched and red shifts. This kind of behavior was observed in CIGS, CZTSSe thin film solar cells. The donor and acceptor energies were determined from the energy of the donor - acceptor pair energy as given by the following relation.

$$E_{DAP} = E_g - E_D - E_A \tag{6.14}$$

Where E_{DAP}, E_g, E_D and E_A are energies of donor-acceptor pair, band gap, donor and acceptors energies, respectively.

In another way using PL is time resolved photoluminescence (TRPL) is used to determine minority carrier life time in the semiconductors. It is widely used method for determination of life time of compound semiconductors such as CIGS, CZTS, CZTSSe and other photovoltaic materials. In this method the time decay of the photo excited carriers are monitored as function of time. A typical PL decay curve is shown in the Figures 6.15 and 6.16.

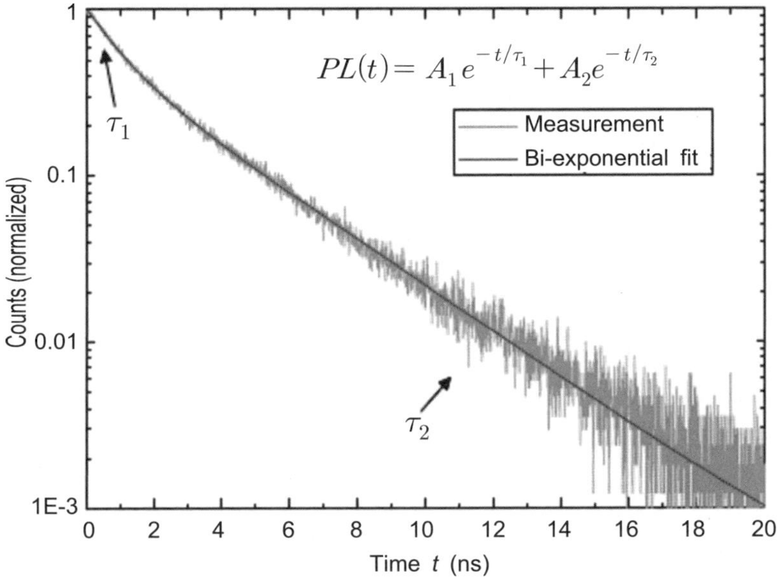

Fig. 6.15. Time resolved photoluminescence of CIGSSe Solar cell [10].

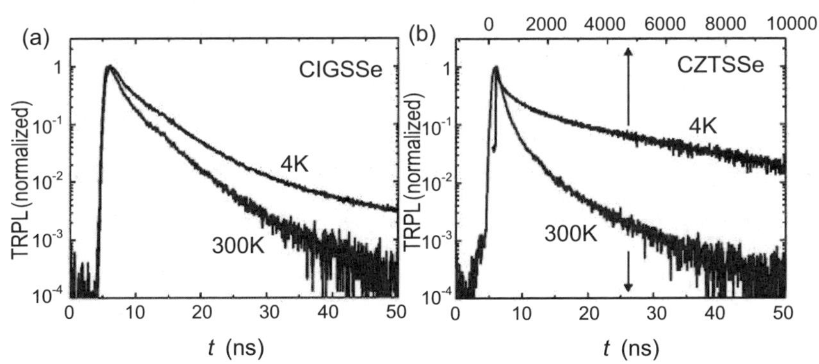

Fig. 6.16. Time resolved photoluminescence of CIGSSe and CZTSSe Solar cells [11].

From the fitting of the decay curve we can estimate the life time of the minority carriers. It is very important tool to evaluate the quality of the chalcogenide photovoltaic materials. For high efficiency CIGS solar cells have about 200 ns whereas the kesterite based CZTSSe solar cells have life time 10 ns. This is also one of the reason kesterite solar cell exhibit lower performances compared to CIGS solar cells. In order to achieve high life time of minority carriers the absorbers must be grown without secondary phases and defects. TRPL is suitable method for determination of life times of as deposited and completed solar cells.

Photoluminescence is also used in imaging of the polycrystalline Si, CIGS, CdTe, CZTSSe thin film solar cells. In polycryalline solar cells the grain boundaries is source of defects which acts as recombination centres in solar cells. Understanding the recombination kinetics and controlling the defects is paramount importance in thin film solar cells. The photoluminescence images of Si and CIGS are shown in Fig. 6.17.

It can be seen from the figure that variation of intensity across sample indicates originating from defects in the absorbers. PL imaging is also used for quality control of solar cells manufacturing due to non-destructive method for investigation of devices.

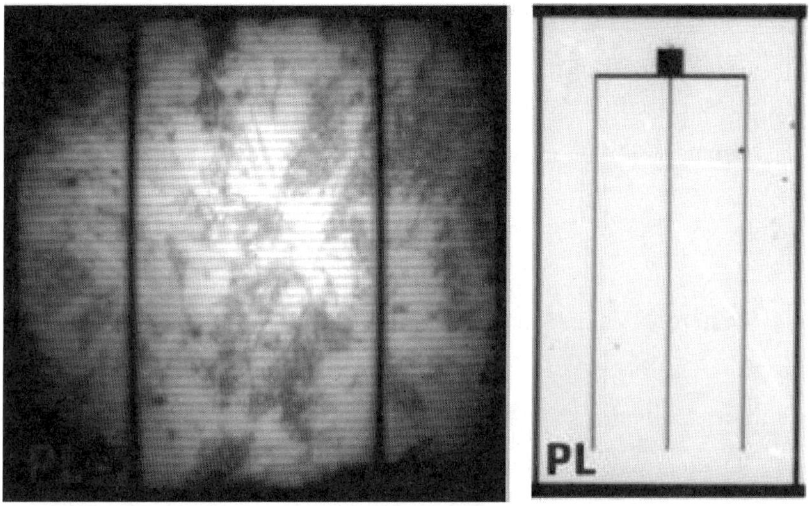

Fig. 6.17. Photoluminescence images of Silicon and CIGS, adapted from [11].

6.8. Capacitance - voltage measurements

Defects in the forbidden band gap affect thin film solar cells. The intrinsic defects such as vacancies, interstitials and anti-sites forms in CIGS, CdTe and CZTSSe materials. The shallow defects have minimal influence on the performance of the devices. However, the deep defects at the mid gap of the semiconductor degrade the performance of the solar cells due enhanced recombination. It is essential to identify these defect energy levels and their distribution to control these defect densities. Typically, capacitance-voltage (C-V) measurements were performed on Schottky and completed p-n junction solar cells to determine the concentration of majority carriers in the bulk of the device, energy levels of interface states at the junction interface. The net carrier concentration can be estimated using the slope of the capacitance $1/C^2$ versus voltage curve from the following equation.

$$N=\left(\frac{2}{qA^2\epsilon\varepsilon_0}\right)\frac{dV}{d}\left(\frac{1}{c^2}\right) \tag{6.15}$$

Where N is the net carrier concentration, q is the electron charge, ε is the permittivity, ε_o is the relative permittivity, A is the area, C is the capacitance and V is the applied voltage.

Deep level transient spectroscopy (DLTS) was used to determine the electrically active defects states at the interface of p-n junction in the solar cells. In this method the charge flows in and out of energy states corresponding to the defects in the forbidden energy band gap. It is used to identify trap states of majority and minority carriers, its activation energies levels and capture cross section.

Recently, another variant of this method was developed it is known as drive level capacitance profiling (DLCP). It is similar to capacitance-voltage measurements with this it is possible to know the carrier concentration as function of depth in the absorber. It is widely used in CIGS, CZTS, CZTSSe solar cells to investigate the deep levels, and non-uniformity in the samples. The measurement is carried out on completed thin film solar cells. During the measurement small alternating current AC signal is superimposed on DC. Using DLCP energy and distribution of defects can be evaluated. In this technique the profiling of defects as function depth can be known. These measurements can be carried out on completed thin film solar cell [12].

6.9. Admittance spectroscopy

Admittance spectroscopy is another capacitance - frequency technique widely used for investigation of thin film solar cells as a function of frequency and temperature. The measurement is carried out on completed thin film solar cells to identify the traps in the forbidden band gap. As we already know that traps in the band gap acts as recombination centres, it is necessary to understand the origin of these traps in order to reduce these defects. The depletion capacitance of the p-n junction solar cell is given by

$$C_{dep} = \frac{\epsilon_r}{W} = \left(\frac{\epsilon_r q N_A}{2 V_{bi}} \right)^{\frac{1}{2}} \tag{6.16}$$

here ε_r dielectric constant of semiconductor, w is the depletion width, q the electronic charge, N_A the acceptor density and V_{bi} is the built-in potential.

The traps in the depletion region increase the junction capacitance at low frequencies and at high temperatures. The increase in capacitance of the depletion region due to majority carrier traps density and its energy positions were measured using admittance spectroscopy. [13, 14]

6.10. Electro luminescence

Electroluminescence is inverse process of solar cell, in which electric current is converted into light. The electroluminescence is widely used technique to understand the defects of semi-conductors. When current applied to a semiconductor junction the generated carriers are recombined radiatively emitting the photons. The recombination is less effective in indirect semiconductors and highly efficient in direct semiconductors. The electroluminescence is used to estimate the band gap; identifying defects, cracks and grain boundary recombination in single and polycrystalline thin film solar cells.

With advance in detectors such as charge coupled devices, electroluminescence has widely used method to evaluate the quality of the absorbers, solar cells and modules during the developing and manufacturing. The image of the absorbers or solar cells shows various contrasts due to presence of structural and electronic defects in the materials, which is used for analysis for electrical losses in solar cells. Fig. 6.18 shows EL image adapted from [6]. All thin film solar cell materials CIGS, CdTe, CZTSSe are polycrystalline in nature. The polycrystalline materials made of grains, grain boundaries which acts as recombination centers and reduces the short circuit current density. And also the non-homogeneities arising from the compositional variation across lateral of the absorbers results in band gap fluctuation which reduces the available open circuit voltage from the materials. These structural non uniformities are well investigated using the EL nd PL imaging techniques to understand the effect of these defects on the performance of the solar cells.

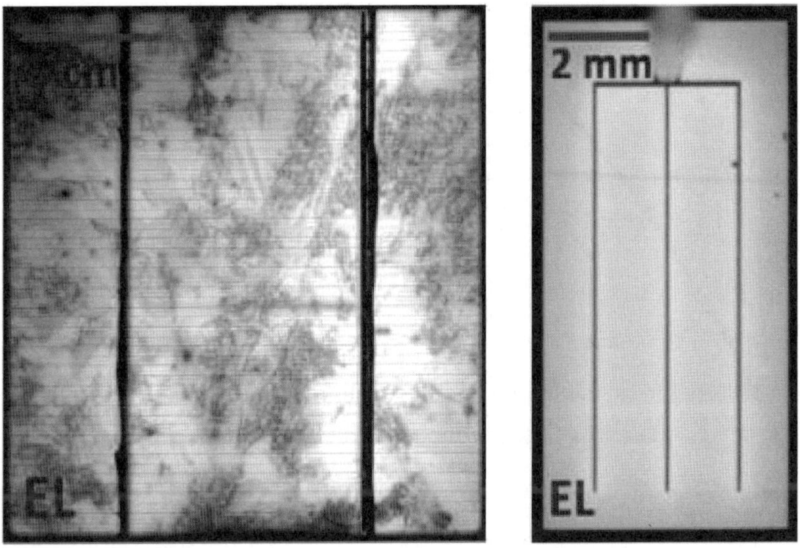

Fig. 6.18. Electroluminescence images of Silicon and CIGS solar cells.

6.11. Simulation of thin film Solar cells

Simulation of solar cells is powerful method for understanding the solar cells. It has advantages that without fabrication of actual solar cells the performance of the devices can understand as a function of various input parameters. There are numerous simulation packages such PC1D, AMPS later WxAMPS and SCAPS were developed for Si and thin film solar cells. PC1D is widely used for simulation of silicon solar cells.

wxAMPS and SCAPS were used for simulation of CIGS, CdTe and CZTS thin film solar cells [15-17].

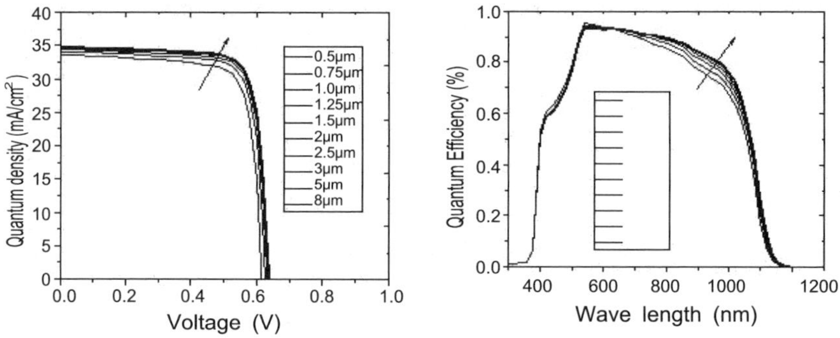

Fig. 6.19. Effect thickness of CIGS absorber on efficiency of CIGS solar cells, simulation I-V curves and EQE curves.

In the following section WxAMPS was used to simulate the CIGS solar cells with variation of CIGS thickness, buffer layer thickness and window layer ZnO thickness to evaluate the performances of the solar cells. Fig. 6.19. shows the simulated current voltage and EQE curves as function of absorber CIGS thickness. As the increase of absorber thickness the J_{sc} increases. The J_{sc} increases due to collection efficiency of carriers, however the collection efficiency decreases as function of thickness.

And also the diffusion length increases with thickness and J_{sc} saturates as the diffusion length increases. the results shows that beyond increase in absorber thickness over 3 μm has no beneficial effect on the performances. This is reason all the high efficiency CIGS solar cell so far achieved with 3 μm thick absorbers.

Fig. 6.20 shows effect of CdS thickness on the performance of the CIGS solar cells. The CdS band gap is 2.42 eV and its corresponding absorption in the CdS layer at wavelengths below ~520 nm. The loss in QE for λ < 500 nm is proportional to the CdS thickness, since carriers generated in CdS is not collected

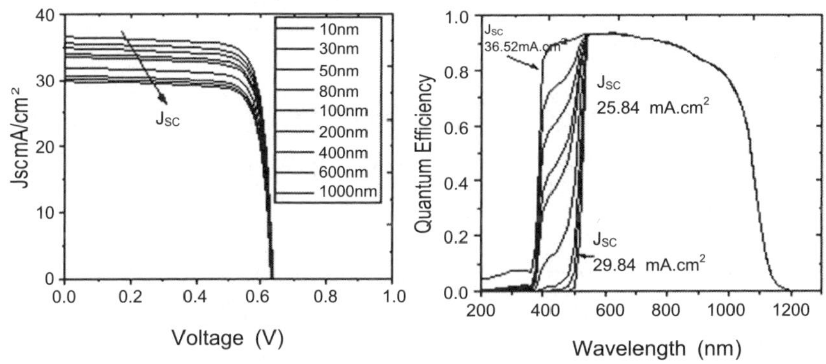

Fig. 6.20. Simulated I-V curves and EQE curves effect of CdS buffer layer thickness on efficiency of CIGS solar cell.

The J_{sc} is difference of J_{sc} ~6 mA/cm^2 for 10 nm to 1,000 nm thick CdS. At 10 nm-thick CdS layer the loss is low, however. In practice, the CdS layer is chosen as 50 nm as optimum thickness for high efficiency and also protection of buffer layer during the window layer deposition by direct current (DC) sputtering. There is no significant variation of ZnO thickness from 200 nm to 1μm on CIGS solar cells as shown in the Fig. 6.21 However, with

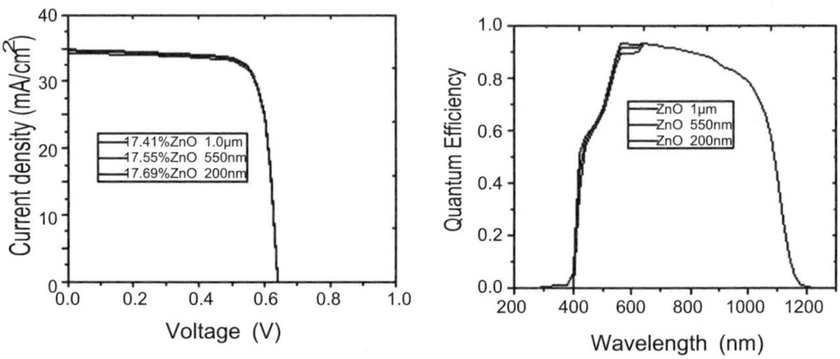

Fig. 6.21. Simulated I-V curves and EQE curves of effect ZnO thickness on efficiency of CIGS solar cell.

Increase in thickness the conductivity increases and transmittance reduces. For large area modules higher thickness is required to reduce resistive losses. The lesser the thickness there is less absorption at blue region of the EQE spectrum which enhances the J_{sc} of the solar cells. The typically used thickness is about 500 nm in CIGS solar cells.

6.11.1. Graded solar cells

CIGS thin film solar cells are fabricated with 'Ga' grading in the absorbers. Taking the advantage of high V_{oc} in 'Ga' rich solar cells and high current with 'In' rich CIGSe thin films it is possible to achieve high efficiency solar cells. However, placing the grading in the devices is important to achieve high performance. The fig. 6.22 shows the 'Ga' grading profile for a five layer CIGS solar cell is used to simulate the performance of the device.

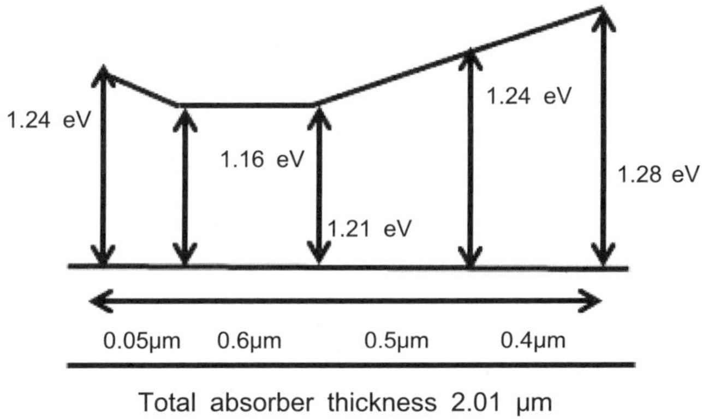

Fig. 6.22. Simulated I-V curves and EQE curves effect of ZnO thickness on efficiency of CIGS solar cell.

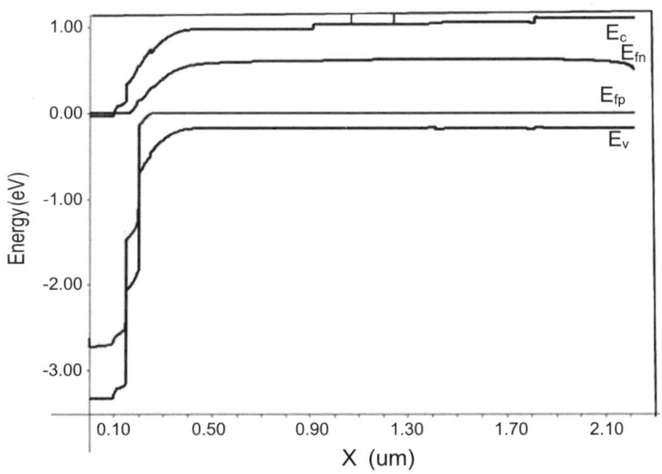

Fig. 6.23. Electronic band structure of CIGS/Cds/ZnO solard cell.

The electronic band structure of the above device is shown in Fig. 6.23. As can be seen from placing absorber layers with various band gap in the devices results creation of band discontinuities which creates a field so that the carriers are drifted to the juction and more carriers are collected. Electric field at the bottom absorbers acts as back sufrace filed similar to one observed in silicon solar cells and reduces recombination at the back side of the devices.

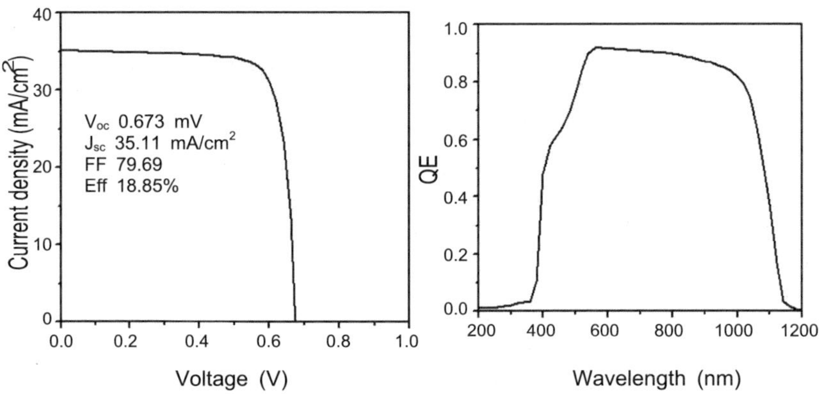

Fig. 6.24. Simulated I-V curves and EQE curves effect ZnO thickness on efficiency of CIGS solar cell.

Fig. 6.24 shows the simulated current-voltage characteristics and quantum efficiency curves of a five layer device. It is clear from the figure that high performance of the devices are achieved with placing a thin CIGS solar cell layer with a band gap 1.24 eV is improves the open circuit voltage. The thin layer wih higher band gap shows degraded performances, it is known that CIGS solar cells show poor performances with gallium ratio above 0.42 with E_g = 1.3 eV. At the back of the solar cell the wide band gap CIGS solar cell acts as back surface field which reduces the recombination at the back there by increasing the efficiency of the solar cells.

Now, many research groups and in production of CIGS solar cells with gallium grading was used in the CIGS solar cells to fabricate high efficiency dvices and large area modules. The simulation is widely used to simulated other CdTe and CZTSSe thin film solar cells because its adaptability to various input parameters [16,17].

6.12. Energy losses in Solar cells

It is clear from the previous sections that a solar cell cannot convert solar radiation completely into electricity. It is necessary to understand the energy loss analysis for optimization of solar cell efficiency. The following factors dictate the available power from solar cells [18].

1. Photons higher than band gap $h\nu > E_g$ are not absorbed instead it heats the solar cell due to thermalization of hot carriers.
2. Low energy photons $h\nu < E_g$ passes through the solar cell without absorption.
3. Collection efficiency η is always less than 1, it means that all carriers are not collected due low L_n and τ_n

4. The available V_{oc} is always less than band gap ($qV_{oc} < E_g$) due location of quasi Fermi levels E_{Fn} and E_{Fp} within the band gap.
5. I^2R losses due to series and shunt resistances R_s, R_{sh} originating from active layers and contacts.
6. Reflection of photons from the surface of the solar cells
7. Area loss due to grid coverage.
8. Unwanted absorption arising from the layers of the solar cells.

Among the above factors 1 and 2 cannot be avoidable. However it can be overcome using graded solar cells and tandem concepts. The remaining losses can be minimized by controlling the defects, doping, improving the minority carrier diffusion length, minority carrier life time τ_n and minimizing the series and shunt resistances. These factors must be controlled in order to achieve high conversion efficiency in crystalline and thin film solar cells. In the next chapter different concepts were explained to overcome the efficiency limits in solar cells.

References

1. Jenny Nelson, "The Physics of Solar cell, Imperial College Press, 2004. (Book).
2. Oki Gunawan, Tayfun Gokmen and David B. Mitzi, "Suns-V_{OC} characteristics of high performance kesterite solar cells, J. Appl. Phys. 116, 084504, 2014 (Journal).
3. Adolf Geotzberger, Joachim Knobloch, Bernhard VoB, "Crystalline Silicon solar cells, John Wiley & Sons., 1998. (Book).
4. S. Hegedus, W.N. Shafarman, "Thin-Film Solar Cells: Device Measurements and Analysis". Prog. Photovoltaics 2004, 12, 155-176 (Journal).
5. James R. Sites, "Device physics of thin film polycrystalline cells ans modules" under Subcontract ADJ-1-30630-06, National Renewable Energy Laboratory, USA (2004). (Report).
6. www.omacom.co.kr, "Quantum efficiency measurement guide" (Web source).
7. Martin A. Green, Keith Emery, Yoshihiro Hishikawa, Wilhelm Warta and Ewan D. Dunlop, "Solar cell efficiency tables (version 47), Prog. Photovolt: Res. Appl. 24, 3-11 (2016), (Journal).
8. I. L. Repins, H. Moutinho, S. G. Choi, A. Kanevce, D. Kuciauskas, P. Dippo, C. L. Beall, J. Carapella, C. DeHart, B. Huang, and S. H. Wei, Indications of short minority-carrier lifetime in kesterite solar cells J. Appl. Phys. 114, 084507 (2013). (Journal)
9. Kong Fai Tai, Talia Gershon, Oki Gunawan, and Cheng Hon Alfred Huan, Examination of electronic structure differences

between CIGSSe and CZTSSe by photoluminescence study, J. Appl. Phys. 117, 235701, 2015 (Journal).

10. Robin Knecht, "Characterization of industrially processed chalcopyrite solar cells with varied absorber composition", Dissertation, University of Oldenberg, (2012).
11. Katherine Zaunbrecher, "Imaging as Characterization Techniques for Thin-Film Cadmium Telluride Photovoltaics", Dissertation, Colorado State University, (2014).
12. Heath, Jennifer T, J. David Cohen, William N. Shafarman. "Bulk and Meta Stable Defects in $CuIn_{1-x}Ga_xSe_2$ Thin Films Using Drive Level Capacitance Profiling. J. Appl. Phys. 95.3 (2004). (Journal).
13. Walter, R. Herberholz, A. C. Müller, and H. W. Schock, "Determination of defect distributions from admittance measurements and application to $Cu(In,Ga)Se_2$ based heterojunctions", J. Appl. Phys. 80 (8) 4441-4420 (1996). (Journal).
14. Habibe BAYHAN, A. Sertap KAVASOGLU, "Admittance and Impedance Spectroscopy on $Cu(In,Ga)Se_2$ Solar Cells", Turk J Phys 27 (2003), 529-535. (Journal).
15. D.A.Clugston and P.A. Basore, "PC1D Version: 32 bit solar cell modeling on personal computers," Proc. 26 th IEEE Photovoaltaic Specialists Conference, Anaheim, pp.207-210, 1997 (Journal).
16. Marc Burgelman, Johan Verschraegen, Stefaan Degrave, Peter Nollet, "Modeling Thin-film PV Devices", Prog. Photovolt: Res. Appl. 143-153 (2004), (Journal).
17. Y. Liu, Y. Sun, and A. Rockett, "A new simulation software of solar cells-wxAMPS", Solar Energy Materials and Solar Cells, 2012.

(Journal).

18. Alan L. Fahrenbruch and Ricahrd H.Bube, “Fundamentals of solar cells”, Academic Press, 1983 (Book).

Chapter 7
Limitations and future directions

7.1. Maximum achievable efficiency of solar cells

According to Shockley-Quisser limit, the maximum achievable limit of solar cells is 30% [1]. This efficiency depends upon the band gap of the absorber material. If band gap is too high, the incident photons due to lower energy will pass without generating electron-hole pair in the absorber compound. On the other hand, if the band gap is too low, incident photons will have excess energy which will be dissipated as heat. Hence the band gap must be optimal in the range of 1-2 eV. Intermediate band gap materials can be employed that absorbs electrons from valance band and transfers to conduction band which can improve the efficiency upto 63%. An electron with high kinetic energy can impact bound charge carrier (impact ionization) and create free electron-hole pairs for conduction which can enhance the performance of solar cell.

7.2. Losses in thin film solar cells

Thin film solar cells have attracted considerable attention in the recent years due to its capacity to satisfy the current energy needs. Quaternary and ternary compound based CZTS and CTS thin films have achieved maximum power conversion efficiency (PCE) upto 12.6% and 6%, respectively [2]. However, the achievable theoretical PCE of the CZTS and CTS solar cells is upto 32.2% and 30%, respectively [2, 3]. Hence as seen, both quaternary and ternary absorber compound based thin film solar cells have scope to improve their PCE upto their theoretical efficiency value. However, there are several limitations to achieve this efficiency due to many factors. Fig. 7.1 shows the losses in solar cell.

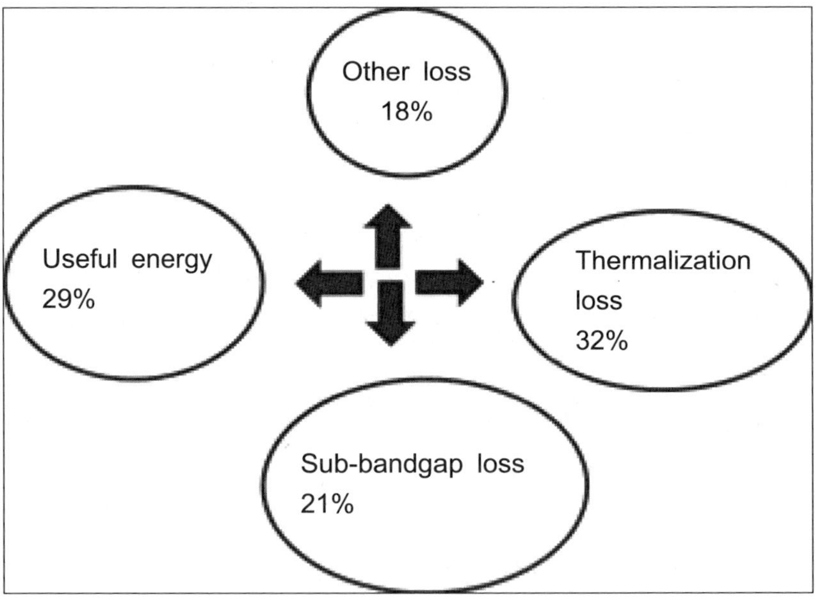

Fig. 7.1. The losses in solar cell.

In case of chemical synthesis of absorber compounds, preparative experimental parameters such as chemical property of solvent, additives, temperature and reaction time significantly affects the quality of absorber compound and hence the PCE [2]. The presence of solvent employed for the chemical synthesis of absorber compound can induce carbon impurities and this greatly affects the efficiency of solar cells. The major limitation in the thin films solar cells is the synthesis of phase pure absorber compound. The secondary phase in the absorber compound creates intrinsic defects and affects the electrical and optical properties which hamper the performance of solar cell.

Other important factors that affect the PCE of solar cells are the p-n junction interface, absorber compound properties and annealing conditions. The maximum possibility of the recombination of the generated electrons and holes are at the p-n junction interface. If the junction is rough and the contact between is non-uniform, recombination occurs which leads to loss of generated charge in the solar cell (Fig. 7.2).

The band alignment match between the p-n junction also affects the performance of solar cells. The ideal 'spike' like structure is favorable as it provides transfer of electrons across the junction interface by thermionic emission and the activation energy and band gap energy are equal. In case of 'cliff' like structure, there is reduction in interface bandgap and reduction in activation energy than band gap [4]. The sulfides in absorber compound forms cliff like structure with CdS (buffer layer) and spike like

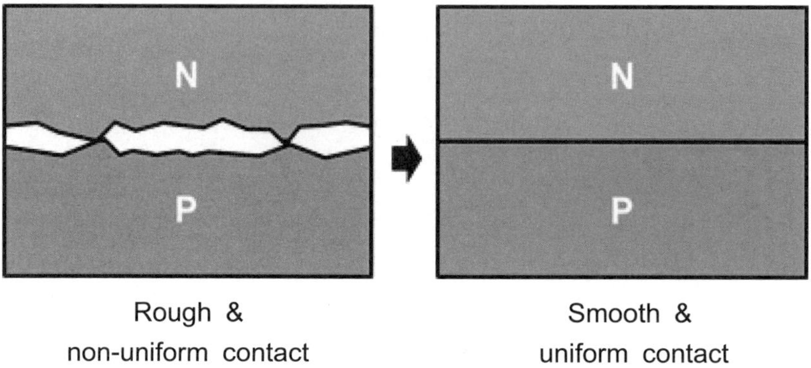

Fig. 7.2. Schematic of contact between p-n junction interfaces.

structure with selenides [5]. Hence, proper selection of n type buffer layer must be done based on the type of absorber compound employed.

Defects, secondary phases and non-stoichiometric composition in the absorber compound also affect the solar cell performance. Control of metal precursor ratio is challenging. Copper rich absorber compound results in formation of copper sulfide secondary phase which causes shunt losses in the solar cell due to their high electrical conductivity. On the other hand copper poor composition in the absorber compound can produce tin rich composition and tin sulfide secondary phases. Hence, composition control is very important for solar cell performance. The loss of tin during annealing treatment in the form of SnS and SnS_2 also affects the composition of the absorber compound. Moreover, tin loss also affects the absorber microstructure resulting in non-compact morphology which causes recombination losses. Hence, tin loss from the absorber compound must be controlled and this can be achieved by using seal tubes for annealing which can suppress the tin loss by the generated pressure of the liberated gas. The presence of MoS_2 layer between the Mo/SLG substrate and absorber compound also affects the performance of solar cells. This layer reduces the minority carrier lifetime and hence the PCE. The layer cannot be totally eliminated, but its thickness can be controlled by Ag coating (20-40 nm) between Mo/SLG and absorber compound (Fig. 7.3) and PCE of solar cell can be enhanced [6].

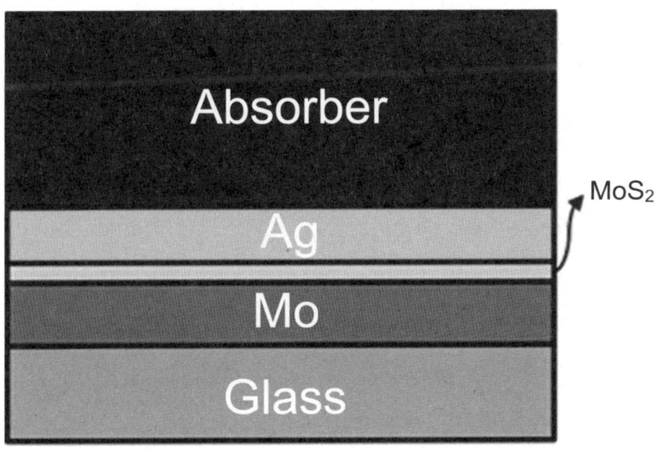

Fig. 7.3. Schematics of MoS_2 layer thickness control by thin Ag coating between Mo substrate and absorber compound.

Doping of alkali metals (Na, K and Li) in the absorber compound has also achieved great attention to enhance PV efficiency. Na doping improves the crystallinity of absorber compound and enhances V_{oc}. Li doping reduces deep defects and also enhances V_{oc} by affecting the electrostatic properties at grain boundaries [7-9]. K doping suppresses tin evaporation and enhances J_{sc}. Hence, alkali metal doping has significant effect on the quality of absorber compound and its PV efficiency. The third level metric losses such as break current loss (J_{sc}), voltage loss (V_{oc}) and fill factor loss (FF) contribute to the performance degradation of the solar cells. The J_{sc} losses are attributed due to reflection, light absorption by buffer layer, glass absorption and deep defects. V_{oc} losses are limited by dominant current transport mechanisms. V_{oc} depends upon recombination and hence V_{oc} can be improved by reduced recombination. The increase in the A-factor and reduction in V_{oc} can reduce FF losses. Low series resistance, shunt conductance and back contact resistance also contribute to low FF loss [10].

The stability of the solar cells is also important. The stability of the solar cells against temperature and moisture is still a challenge. The V_{oc} and FF are significantly reduced due to moisture and hence the performance of the solar cell is affected. This problem can be partially tackled by encapsulation, frame sealing and glass to glass configuration with active barrier as gasket. The stability of the solar cells must also be studied to ensure longer performance stability.

The morphology of the absorber must be compact with highly crystallized grains Fig. 7.4((a, b)). Smaller grain will have more grain boundary density and lead to more recombination causing reduction in the carrier life time. Therefore the grain size must be controlled and is possible by annealing treatments. Hence, the sulfurization conditions must be optimized.

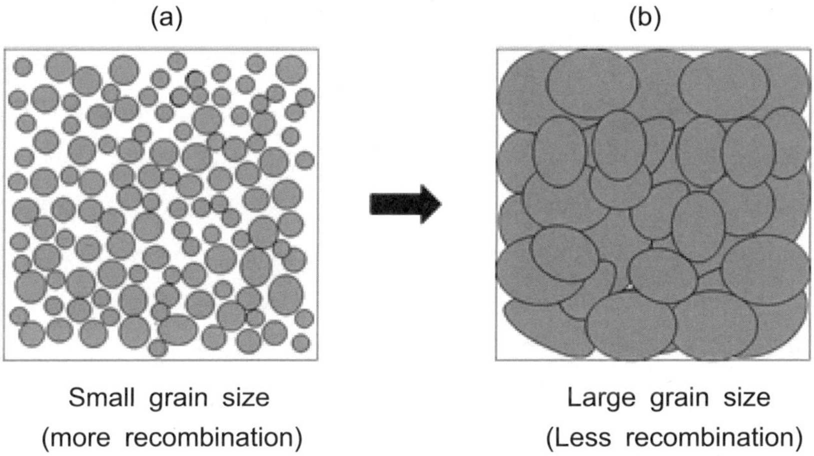

Fig. 7.4. Schematic of (a) small grain size and (b) large grain size.

Other factors such as non-absorption, thermalization, reflection, transmission and area loss also contribute to the efficiency loss in solar cells. Photons with energy lower than the band gap energy of the absorber material is transmitted through the cell without absorption and hence do not contribute to the power output of cell. Thermalization loss occurs when the energy of the incident photons is greater than the band gap of the absorber material. This high energy increases the temperature of the solar cell which further increases the reverse saturation current and decreases open circuit voltage due to increase in carries concentration and their diffusion length. Reflection losses occur due to the reflection of light from the top surface and back contact surface. Finite thickness of the absorber layer leads to transmission loss and this effect is more in compounds having low absorption coefficient. The metal grid design on the top surface of solar cell leads to loss and it restricts the light from absorbing. These optical losses can be reduced by employing antireflective coating (ARC) on the top surface. Texturing on the top surface can be done to reflect light in proper manner and avoid reflection losses. Surface and bulk recombination at semiconductor contact and recombination in depletion region leads to collection losses which significantly affect the open circuit voltage [1]. The presence of impurities, defects and incomplete chemical bond in semiconductor cause reduction in photocurrent. Series resistance occurs due to bulk resistance of the absorber compound and metal contact between semiconductor and metal. Proper design of top contact can help to reduce the resistance. The shunt resistance is due to the leakage across the p-n junction and impurities. Both series and shunt resistance reduce the solar cell efficiency and hence the resistance must be low as possible.

7.3. Methods to enhance the performance of thin film solar cells

7.3.1. Antireflection coating (ARC)

ARC is usually applied to reduce the light reflection from the top surface of solar cell. The layer is transparent and quarter wavelength thick and is made up of dielectric material. The light reflected from the ARC is 180° out of phase with the reflected light from the semiconductor absorber material [1, 11]. This results in destructive interference making the resultant reflected energy zero. The structure of ARC is shown in Fig. 7.5. The ARC is made up of MgF_2 and Si_3N_4. ITO and ZnO/pores silicon layer can also be used as ARC.

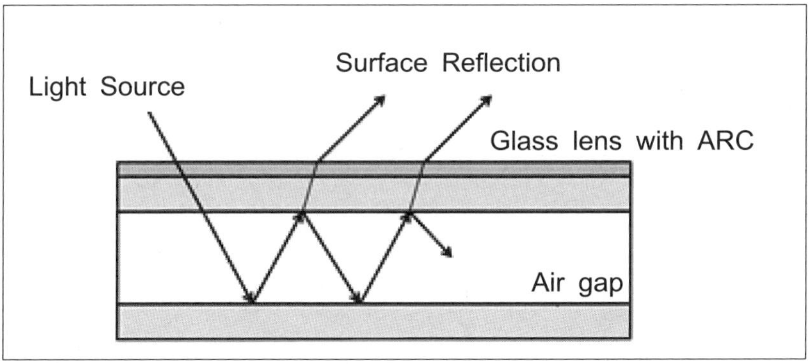

Fig. 7.5. The schematic diagram of antireflection coating (ARC).

7.3.2. Surface texturing

Surface texturing is usually done to reduce light reflection by making the surface rough. Texturing the surface causes light to bounce back to the surface and reduces light reflection from the surface. Random and uneven texturing is more effective than flat surface. Inverted, pyramidal and upright texturing is more effective. These geometrical texturing is not always suitable for all types of solar cells. Especially for thin film solar cells, as the absorber thickness is low, texturing can lead to low absorption of light leading to minority carrier recombination. Pyramidal texturing is suitable for monocrystalline solar cells as they enhance the efficiency of solar cells by reducing bulk recombination even at lower absorber thickness [12]. Fig. 7.6 shows the image of texturing.

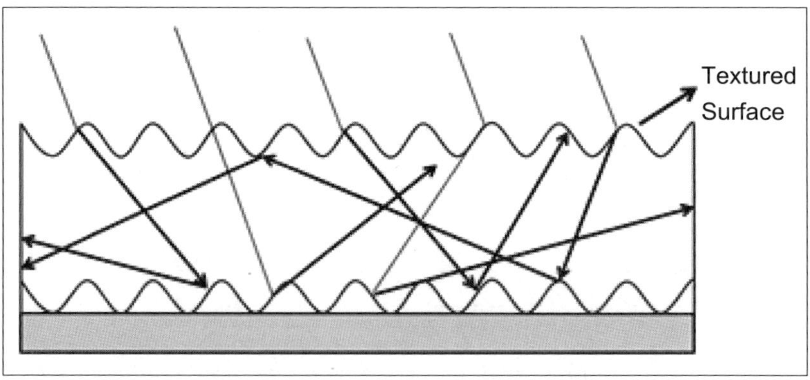

Fig. 7.6. The schematic diagram of texturing.

7.3.3. Back surface field (Braggs reflector)

The generated light photons do not stay trapped inside the solar cell for long time in the condition when the absorber layer thickness is less than the wavelength of light photons. Increasing the optical path length is desirable and is possible to achieve by using ARC and surface texturing. The optical path length of the photons is equal to the thickness of the absorber material when the light is perpendicular to the device. In that case, some photons may not get trapped in the device. When light falls at some inclination on the surface of device, the optical path length of the photons increases and light gets reflected and trapped inside for a longer time reducing the recombination of photons at back surface [13]. Heavily doped Al and Ag particles are used for making back surface. In solar cell configuration, Braggs reflector is deposited on the absorber layer.

7.3.4. Surface passivation

The presence of dangling bonds, impurities and surface discontinuities are the reasons for the recombination at the front and back surface of the solar cell device which causes reduction in the cell efficiency. To reduce recombination, the surface is passivated by using dielectric surface such as Si, SiO_2 and Si_3N_4. Passivation can also be done using chemical treatment of passivizing oxide [14].

7.3.5. Absorber material thickness

The efficiency of the solar cell strongly depends upon the thickness of the absorber material. Photons with energy higher than band gap and the photons with energy lower than band gap are lost in the form of radiation due to the thickness of the absorber material. The thickness of the absorber compound must be equal to the absorption length or wavelength of the photon [1].

In case of thin film solar cells, the thickness of the absorber material lies in the range of 1-1.5 μm.

7.3.6. Metallic nanoparticles (plasmonic effect)

The thickness of the absorber material is important in thin film solar cell. Higher the thickness, higher is the absorption. But this thickness can be maintained low as possible without affecting the absorption of the light. Metallic nanoparticles (Ag) can be employed on the top layer of active material to scatter the incident light. When light falls on the surface of the solar cell, it is scattered in different directions. These nanoparticles can scatter the light into high index material (buffer layer) to enhance coupling with semiconductor and reduce reflection losses [1]. Nanoparticles couple the plasmonic near field to semiconductor, and thus increase the effective absorption. Fig. 7.7 shows the scattering of light from metal nanoparticles at the surface of solar cell.

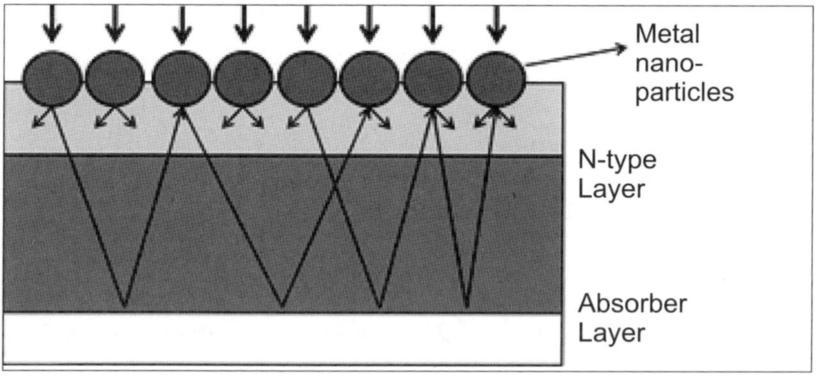

Fig. 7.7. The schematic diagram of scattering of light from metal nanoparticles at the surface of solar cell.

7.3.7. Electron reflector

Charge carrier density and electron lifetime can improve open circuit voltage and conversion efficiency of solar cell. Under forward bias condition, electron reflector will reflect minority carrier electrons and reduce back surface recombination. This electron reflector can be formed by adding a layer of p-type material with large band gap material at the back surface of absorber compound. This method is quite easy and cheap.

7.3.8. Solar tracker

Solar trackers can be used to track the generated power by controlling current and voltage and adjust the load in such a condition that always maximum power is transferred irrespective of the lightning conditions. Solar tracker ensures the sun radiation to fall perpendicular on the device surface to ensure the optical path length of the photons equal to the thickness of the absorber material.

7.3.9. Solar concentrators

The device performance increases with the increase in number of photogenerated carries. Solar concentrators can increase the light intensity and hence the photon carriers. A typical concentrator uses a light intensity of 4-400 times the sun.

7.3.10. Downshifting

This is conversion process in which high energy UV and blue phonons are down converted into visible spectrum where the quantum efficiency value is 100 %.

7.3.11. Down and up conversion

Down conversion process is a process of splitting one high energy photon into two photons that sufficiently have energy to create electron-hole pair. In up conversion process, two low energy photons are combined that has enough energy to create electron-hole pair.

7.4. Future developments in solar cells

7.4.1. Tandem solar cells

The present thin film solar cell efficiency is limited due to the narrow absorption bands of the active layers. Tandem solar cells were developed to fully utilize the solar spectrum and enhance the performance of solar cells. Tandem solar cells are prepared by combing several junctions in one solar cell. As seen in Fig. 7.8, the solar cell is stacked where two band gap active layers are connected each other by recombination layer. The top layers

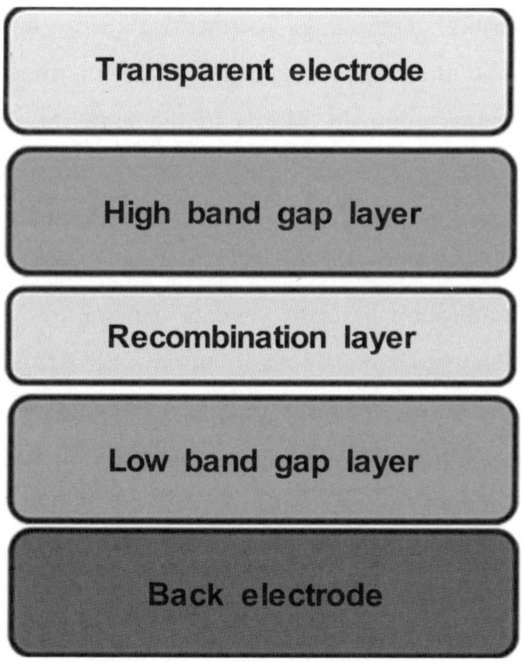

Fig. 7.8. The schematic diagram of tandem solar cell.

absorbs high energy photons and bottom layer absorbs low energy photons which causes to utilize entire solar spectrum and increase in cell performance due to reduced thermalization losses which is proportional with the energetic difference between a given absorbed photon and the band-gap of the absorbing semiconductor [15]. The individual cell can be connected in series and parallel configuration. Series connected cells are simpler to fabricate but the current is the same though each cell so this constrains the band gaps that can be used.

7.4.2. Intermediate band gap solar cells

As the name suggests, an intermediate energy level is formed between valance and conduction bands. The intermediate band is formed by doping/ impurity atoms to the material. These doped atoms modify the energy level and create additional intermediate energy band within the band gap of the material. This allows the photons with energy less than the bandgap to excite an electron to conduction band to valance band. Fig. 7.9 shows the intermediate energy band in the material. Elements such as Ti and Cr can be doped to form intermediate energy level. These cells are also called as multiband solar cells.

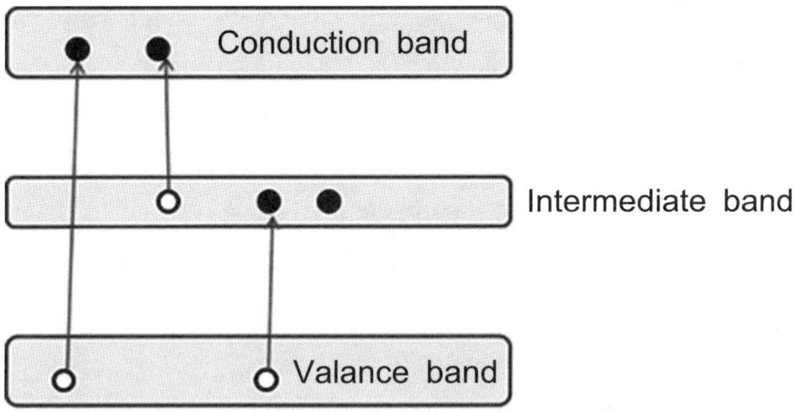

Fig. 7.9. The schematic diagram of intermediate energy band in the material.

The demand for solar cells in the near future will be increased. Understanding the current drawback of the solar cells and implementing the solutions for the same will help to enhance the performance. The improvement in the performance of solar cells can drive attention to sustainable, clean and free solar energy as an alternative to fossil fuels.

7.4.3. Thin film full spectrum solar cells

The theoretical efficiency of thin film solar cells is nearly 35 %. To achieve this efficiency, the absorber compound must be defect free, should possess high quality. The p-n junction must also be of good quality. As for ideal conditions, this efficiency is achievable. However it is highly difficult and challenging to reach this high theoretical efficiency value. This current challenge can be solved by quintuple-junction solar cells which employs light absorbing material stacking layers with different band gap ranging from 0.6 to 2.2 eV. As seen in Fig. 7.10, the first cell can be made of Si thin film with band gap of 2 eV. Second cell can be prepared by using amorphous Si. Third cell having band gap of 1.4 eV can be used with a-SiGe and Si quantum dots. Fourth cell can be fabricated using μc-Si. The fifth cell can be fabricated using μc-SiGE [16].

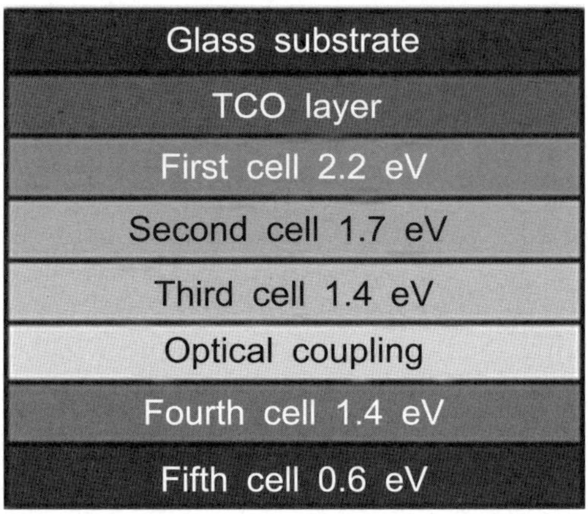

Fig. 7.10. Schematics of quintuple-junction solar cells.

7.4.4. Silicon wire PV solar cells

It consists of using an array of microscopic silicon wires on a plastic substrate. The design uses only 1% to 2% of the silicon used in crystalline silicon solar cell and has the added advantage that the plastic substrate is flexible and it can be applied on non-flat surfaces [17].

7.4.5. 60% efficiency solar cells

Production of solar cells using quantum dots could double the efficiency levels currently possible and reduce costs [17].

7.4.6. Self-Replicating high efficiency solar sells

Researchers have developed solar cells that self-replicate like chloroplasts in plants - achieving initial efficiencies of 40% [17].

7.4.7. Electrowetting to increase efficiency

Using an electric charge to change the shape of a fluid helps concentrate light onto the photovoltaic cell and can increase electrical output [17].

7.4.8. NOC on solar cells for protection

NOC on solar is a hydrophobic coating that can be applied on new or existing solar panels that helps make them self-cleaning and increases electricity production [17].

The three most viable thin-film photovoltaic technologies - cadmium telluride (CdTe), copper-indium gallium (di) selenide (CIGS), and amorphous silicon (a-Si) - continue to mature and grow technologically and in market stature.

But apart from the dominance shown by CdTe leader First Solar, the rest of the TFPV manufacturers have had a fairly difficult time making significant commercial inroads as the price of mainstream crystalline-silicon modules plummeted over the past couple of years. Other factors delaying the long-predicted age of thin film include bankability challenges and difficulties in reducing production and system costs [17].

References

1. M. Dhankhar, O. P. Singh, V.N. Singh, Renewable and Sustainable Energy Reviews 40, 214-223 (2014).

2. A. C. Lokhande, K.V. Gurav, Eunjin Jo, C .D. Lokhande, Jin Hyeok Kim, J. Alloys Compd. 656 295-310 (2016).

3. W. Shockley, HJ. Queisser, J Appl Phys; 32: 510-9 (1961).

4. S. Sato, H. Sumi, G. Shi, and M. Sugiyama, Phys. Status Solidi C 12, 757-760 (2015).

5. S. Siebentritt, Thin Solid Films 535, 1-4 (2013).

6. H. Cui, C. Y. Lee, W. Li, X. Liu, X. Wen, and X. Hao, Int. J. Photoenergy, 2015, 1-9(2015).

7. Y. Yang, X. Kang, L. Huang, S. Wei and D. Pan, J. Phys. Chem. C 119, (2015) 22797-22802.

8. H. Xin, S. M. Vorpahl, A. D. Collord, I. L. Braly, A. R. Uhl, B. W. Krueger, D. S. Gingerbc and H. W. Hillhouse, Phys. Chem. Chem. Phys., 17, 23859 (2015).

9. Z. Tong, C. Yan, Z. Su, F. Zeng, J. Yang, Y. Li, L. Jiang, Y. Laiand, F. Liu, Appl. Phys. Lett. 105, 223903 (2014).

10. S.H. Demtsu and J.R. Sites, 0-7803-8707-4/05/$20.00 2005 IEEE

11. S. Chattopadhyay, YF. Huang, YJ. Jen, A. Ganguly, KH. Chen, LC. Chen. Mater Sci Eng R 69, 1-35 (2010).

12. A. Goetzberger. Optical confinement in thin Si-solar cells by diffuse back reflectors. In: Proceedings of the 15th IEEE photovoltaic specialists conference, Kissimmee; 1981. p.867.

13. A. Luque, S. Hegedus. Handbook of photovoltaic science and engineering. 2nd ed. New York, USA: John Wiley and Sons Ltd.; 2011.

14. R. Brendel. Thin film crystalline silicon solar cell. Germany: Wiley-VCH Gmbh and Co. KGaA; 2003 (chapter 3).

15. Hirst et. al., Progress in Photovoltaics: Research and Applications 2011 10.1002/pip.1024

16. M. Konagai, Current Status of Thin-Film Solar Cells and Future Prospects, Active-Matrix Flatpanel Displays and Devices (AM-FPD), 2013 Twentieth International Workshop on, IEEE, Japan.

17. http://www.solar-facts-and-advice.com/thin-film.html.

Index